Maximum Entropy in Actio

Maximum Entropy in Action

A COLLECTION OF EXPOSITORY ESSAYS

Edited by

BRIAN BUCK
and
VINCENT A. MACAULAY

Department of Theoretical Physics,
University of Oxford

CLARENDON PRESS · OXFORD
1991

Oxford University Press, Walton Street, Oxford OX2 6DP
Oxford New York Toronto
Delhi Bombay Calcutta Madras Karachi
Petaling Jaya Singapore Hong Kong Tokyo
Nairobi Dar es Salaam Cape Town
Melbourne Auckland
and associated companies in
Berlin Ibadan

Oxford is a trade mark of Oxford University Press

Published in the United States
by Oxford University Press, New York

British Library Cataloguing in Publication Data
A catalogue record for this book
is available from the British Library.

Library of Congress Cataloging in Publication Data
(Data available)
ISBN 0–19–853941–X
ISBN 0–19–853963–0 (pbk)

This book was typeset by the editors in LaTeX on the phototypesetter of the University of London Computer Centre

Printed in Great Britain by Bookcraft (Bath) Ltd
Midsomer Norton, Avon

Preface

This collection of introductory articles on maximum entropy and Bayesian methods grew out of a series of lectures given in the Physics Department of the University of Oxford in the summer of 1989. They were arranged by the editors in association with the Physical Sciences Faculty Board to be interdisciplinary in nature, and drew a large audience from across the spectrum of the sciences.

We were encouraged to try to make them available in printed form and we are grateful to Professors Roger Elliott and Chris Llewellyn Smith for bringing the notion of such a book to the attention of the University Press.

To the editorial team at OUP, who guided us through the new territory of book production with constant patience, many thanks.

The editors 'set the type' themselves from variously word-processed scripts sent by the authors, using Leslie Lamport's user-friendly interface LaTeX to Donald Knuth's TeX program with a 'style' supplied by OUP's house 'TeXnician', for which we thank him. The final high resolution bromides were typeset on the phototypesetter at the University of London Computer Centre by the editors.

The splendid diagrams were mostly prepared in the Drawing Office of the Department of Nuclear and Astrophysics here in Oxford by Irmgard Smith for whose conscientious work on some difficult material we are very grateful. The remainder of the diagrams were supplied by the authors themselves.

Oxford
December 1990

B. B.
V. A. M.

Table of contents

Editors' introduction

Dedicated readers of research journals have probably become aware in recent years of an increasing number of articles mentioning, or even making use of, an intriguingly named new method, and they may have wondered if there was something in it for them. This new technique of maximum entropy is indeed very powerful and has now found application in both practical and theoretical studies ranging from image enhancement to nuclear physics, from statistical mechanics to economics. The reason for this wide application is not that the method is a new physical theory of anything, but that it provides a much needed extension of the established principles of rational inference in the sciences (and possibly elsewhere).

One impediment to the even more general use of such a valuable development is that connected accounts of the method and its implementation are difficult to find in any of the more widely available books and periodicals. It is also regrettably common to overlook mention of the pitfalls. The easily predictable result is that many would-be users have rejected the idea out of hand or have totally misunderstood its purpose. The intention is basically to give a way of extracting the most convincing conclusions implied by given data and any prior knowledge of the circumstances. It is not a magic black box guaranteed to compensate for inadequate data or to rescue badly designed experiments. The method has its roots in probability theory, which has long been recognized as the only consistent way to reason in the face of uncertainty, and it is in fact a modern enrichment of that ancient art of conjecture.

We, the editors of this volume, have been interested for some time in the problem of inversion, of going from incomplete and noisy data to a description of the underlying physical system, and it occurred to us that the principle of maximum entropy might be of help in its solution. In view of the difficulties mentioned above it also seemed to us that it would be useful to arrange a series of personal tutorials on the subject by the active practitioners living in the UK. By disguising these private tutorials as interdisciplinary research seminars, among the first ever to be scheduled at Oxford, we were able to enjoy some excellent instructional sessions. Since the lectures aroused a great deal of interest throughout the University it further occurred to us that published versions of the talks would perhaps stimulate more people to explore the possibilities for themselves.

In this introduction we give a very brief overview of the origins and uses of the idea of maximum entropy in order to set the scene for the detailed expositions to follow. But first we should say that the original choice of the word 'entropy' was probably a mistake, though one that it is now difficult to rectify. The word seems, for no very obvious reason, to inspire deep emotions, especially among those who do not wish the concept of entropy in thermal physics to be identified with the purely mental construct that we are talking about (actually, the two concepts *are* related as we shall see). The word was suggested to Claude Shannon by John von Neumann to denote missing information; but a better choice, and less emotive, would have been the word 'uncertainty'.

Probability theory comes in two parts, one of which is not dealt with at all satisfactorily in the standard texts. The part that everybody agrees with in practice, though not always by using the same justification, has to do with the manipulation and combination of probabilities and the rules are given, essentially completely, by well-known sum and product expressions. The first states that, on any evidence which is not self-contradictory, the probability that some proposition is true, and the probability that the denial of the same assertion is true, add up to unity. The second one tells how to break up the probability of the truth of two propositions, asserted jointly, as the product of the probability that one of them is true, given the other, with the probability that the latter is true. The imposing edifice of the modern theory can be erected on this very slender basis. What is omitted from the rules is how to assign actual numerical values to the probabilities in the first place so that the formalism can lead to useful results. The task of the second part of the theory is to find systematic ways of assigning these numbers, which can include, and go beyond, the usual appeals to statistical information on relative frequencies or to enumeration of possible outcomes which are judged to be equally likely.

At least one component of the missing general method seems to be provided by the *principle of maximum entropy* or, as we would prefer to say, *maximum uncertainty*. The evolution of this idea stems from the work of Claude Shannon in communication technology. He envisaged that it should be possible to attach to a probability distribution a single number measuring the total amount of uncertainty represented by that distribution. In short, we would feel less uncertain about the real state of affairs, if only a few possibilities out of many were at all likely, than if we had to take seriously a lot more of the possibilities, each with some appreciable probability of being true. By using an argument of consistency (two routes to a measure should give the same answer), a requirement of continuity and a plausible rule for the combination of uncertainties, Shannon arrived at an effectively unique expression for the desired measure, at least for the commonly considered case of propositions which are exhaustive and mutually exclusive in a given situation. These last conditions entail that only one of the proposed assertions is true, though of course we are uncertain which one. This is why we need to use probability theory. The result can be generalized in various directions, for example, in order to apply it to exhaustive but non-exclusive propositions, but we shall consider here a different line of development.

In the simplest case, we have N propositions labelled by $n = 1, 2, \ldots, N$, exactly one of which is true, and by some means we have assigned to them a set of probabilities p_n which encode all our knowledge relevant to choosing one rather than another. The amount of uncertainty in this situation is then, according to Shannon, represented by the formula

$$S = -k \sum_{n=1}^{N} p_n \ln p_n,$$

in which k is an arbitrary positive constant. With suitable choice of k, the result can be interpreted as an estimate of the number of questions (having yes/no answers) which would be needed to isolate the true proposition. In particular, most questions would be needed when all $p_n = 1/N$, i.e., for the uniform distribution of probabilities. Hence the result does seem a reasonable embodiment of the qualitative idea of uncertainty. This expression for the 'entropy' of a probability distribution was used by Shannon in the proof of his fundamental theorem on the efficient coding of messages for transmission over noisy communication channels. The surprising conclusion is that transmission can be made essentially error-free.

The above formula had also appeared long before in physics. It is identical in form to Planck's expression for thermodynamic entropy, in which the probabilities refer to the possible occurrence of various microstates of a macroscopic body and, of course, the quantity S is taken to be a down-to-earth physical property of a body, not a mere summary of our knowledge or uncertainty about the state of the system. Still, it was striking that exactly the same mathematical expression had appeared in a completely different context and it should perhaps have signalled immediately that there was some concept in common. The great contribution of E. T. Jaynes, some years later, was to point out the underlying connection. His idea was that in both examples the formula did indeed reflect in a quantitative way the total amount of uncertainty remaining after all relevant information about some situation had been taken into account. Even more importantly, he proposed that it could be made the basis for a new method of assigning probabilities.

Jaynes' principle, as it is now called, uses *testable* information, that is, propositions which are relevant to a probability distribution and whose truth can be checked when that distribution has been assigned. An example is data on the value of a measured quantity which could be taken as an average over a probability distribution of various outcomes. The principle then states that the best choice of probabilities is that for which the uncertainty number is maximized, subject to the constraints implied by the testable information. Maximizing uncertainty, while ensuring that known data are reproduced, definitely corresponds well with intuitive ideas of honest probability assignment. This variational principle has several virtues, not the least of which is that for common types of testable information it can be proved that the maximizing distribution is unique. No possibility is assigned zero probability unless the data explicitly require it and in fact the resulting distribution is as spread out among the possibilities as is compatible with the known constraints. If, indeed, there is no information other than that the propositions are exhaustive and mutually exclusive (so that the probabilities sum to unity), then the principle of maximum uncertainty yields the uniform distribution, which is in pleasing agreement with common sense.

The expression for uncertainty is easily extended to cover a countable infinity of possibilities and with this form Jaynes showed that his principle gives a convincing basis for the canonical and grand canonical distributions of statistical mechanics. These refer to the probabilities of occurrence of energy eigenstates of a physical system at equilibrium and their assignment from information theory assuming fixed mean energy and particle number shows very clearly that their form does not depend on physics, but rather that they represent our knowledge of the system. Even more striking is the result that if k is chosen as the Boltzmann constant then the thermodynamic entropy and the maximized uncertainty function are numerically equal. Many features of thermodynamics are then seen in an entirely new light and the way is clear to lay solid foundations for a theory of non-equilibrium processes. These results are far-reaching and illuminating, but the real point of the work of Jaynes is the new method of inference, with its numerous potential applications in other fields.

For many such applications the uncertainty formula needs to be modified so that assignment of continuous probability density distributions can be handled. It has now been shown quite convincingly by several authors that the correct form for the uncertainty of a continuous distribution $p(x)$ is

$$S = -\int p(x) \ln \left(\frac{p(x)}{m(x)} \right) \mathrm{d}x,$$

where $m(x)$ is a function determined by the exact nature of the problem. Maximization of this uncertainty shows, for example, that when it is sensible to choose $m(x)$ a constant, and the mean value and variance of a quantity are specified, then the maximally non-committal density is the normal or Gaussian distribution. A further natural generalization is to probabilities in function spaces. Such an extension will always be needed if we are to have a proper rationale for tackling inverse problems. For it is required to construct, from noisy and discrete data and background information on the signal, the probabilities of different functional forms for that signal. The finding of effective ways of applying the principle of maximum uncertainty in function spaces is a topic of current research.

Pending a natural solution of inverse problems along the lines sketched out above, there has evolved an alternative scheme for the reconstruction of signal sources which take the form of positive density distributions. Examples are light intensities of images, particle number densities and spectra of many kinds, all viewed through some recording instrument which may change the original signal to another physical form, discretize it and simultaneously introduce error. This alternative algorithm was pioneered by Drs Gull and Daniell and is also known as the *method of maximum entropy*, but it is not an obvious continuation of the ideas outlined so far. It is still a variational principle and it is almost certainly related to the earlier method, though the connections are not yet completely elucidated.

The main strategy is to imagine the building up of the investigated positive density by placing numerous small quanta into a finite number of cells, so that the density distribution is reasonably well approximated by specifying the numbers of quanta in the various boxes. It is then assumed that the best distribution is that which can be made by the above process in the greatest number of ways while still agreeing with the known data according to some criterion. Thus the method should perhaps be called the *principle of greatest multiplicity* rather than of *maximum entropy*. The final form of the procedure involves the maximization, under data constraints, of the so-called *configurational entropy* of the density distribution. This latter quantity has the same form as the information theory entropy, but expressed in terms of the proportions of the quanta in the cells rather than involving probabilities. Hence the actual procedure followed looks very much like the generation of a probability distribution by means of Jaynes' principle, though the object produced is a physical density function. The method shares some of the intuitively desirable properties of the idea of maximizing uncertainty. In particular, the deduced density is as spread out or uniform as is possible while remaining compatible with the data and it does not contain any feature for which there is no evidence in the data. Furthermore, Dr Skilling has shown, using reasonable axioms, that the positive density of greatest multiplicity is also the most probable one among all those which agree with observation.

The various ideas discussed here are clearly interrelated and we expect that future researches will converge on some generally acceptable philosophy for attacking the difficult and inescapable problems of probabilistic inference. This book gives examples of the present state of the art and is organized as follows. The first two chapters describe the rationale of the maximum entropy method, while Chapters 3–5 explore some applications. The next two chapters then make plain the interpretation of thermodynamic entropy in terms of maximized uncertainty, and the relation of both to the Bayesian probability theory. Finally, an alternative view of these things is given in the context of crystallography.

In more detail, *Of maps and monkeys* gives an overview of the different kinds of tasks that a unified approach to data handling is required to deal with. The only consistent calculus for this process of *inference* is that of probability theory as championed by Laplace, Jeffreys, Cox and Jaynes. Dr Daniell demonstrates that in many problems the amount of data available is extremely small in comparison with the 'size' of the image we are trying to reconstruct from it. As a result, it is not satisfactory to ignore the variation over different images of the *prior probability distribution*—an encoding of what is known about the real image before we consider the current data—and to rely only on the information carried by the data (via the *likelihood*). The intuitively appealing *monkey argument* is used to generate the entropic prior on positive, additive images, a prior which favours reconstructions as uniform as possible. Mathematical complexity is avoided and the informal style of the original talk has been deliberately retained.

In *Fundamentals of MaxEnt in data analysis*, a more sophisticated argument is supplied to justify the status of the entropic prior. This is important: for although the results provided by using MaxEnt speak for themselves, as will be seen later, it is central to the Bayesian outlook that methods follow from plausible axioms in a logically correct fashion; we want to eliminate *adhoc*kery from our procedures. Central to this new derivation is a requirement that our inferences should depend only on objects which Dr Skilling has called *observables*: integrals linear in the image. This is not an easy chapter, but out of the mathematics comes an extension of MaxEnt: the ability to introduce a *preblur*. The traditional entropy prior assumes no correlations between the pixels of an image. However this has never really been satisfactory since images are in fact almost always correlated. The new formalism allows correlations to develop in the image if the data contains evidence for them. This is an exciting new area and we anticipate that future developments of these methods will involve the incorporation of more specific accumulated experience about the nature of particular sorts of images.

The effort in the applications chapters has been to give enough details of the particular field so that the problems to which MaxEnt has been applied are clear in principle. *Maximum entropy and nuclear magnetic resonance* gives a brief account of modern time-domain NMR before considering in detail the pros and cons of using MaxEnt to recover spectra. The case for the use of MaxEnt in this sort of spectroscopy is still somewhat controversial. In general, the spectrum is not a positive, additive distribution: indeed it is not even real. So the last word has not yet been said on Bayesian techniques for the analysis of such signals. Dr Hore presents an alternative formulation, due to Dr Daniell and himself, more in the spirit of the statistical mechanics which was the origin of the use of entropy in inference. Their entropy is defined on a quantum-mechanical density operator and is a genuine encoding of uncertainty.

In Chapter 4, we again look at spectroscopy, this time Raman as well as NMR. Some impressive reconstructions are displayed, which should convince people that there is something important going on here. In a final section, the disaster which results from using *ad hoc* forms of 'entropy' is displayed.

In *Maximum entropy and plasma physics*, we see MaxEnt being used routinely and skilfully as part of the toolkit of a physicist confronted with a variety of the diagnostic problems which arise in the study of confined plasmas. Here the problems of noisy and sparse data can be quite acute. A useful technique for tuning the values of imperfectly known parameters is described.

At this point we make a couple of asides, which the neophyte might wish to skip until he is more familiar with the material. Firstly, we remark on a comment which is common to several of the chapters saying something to the effect that certain physical distributions that are positive (be they of charge, mass, spectral intensity or whatever) 'can be regarded as probability distributions'. As we have said earlier in this introduction, it appears to us that this remark is rather confusing. It seems to be made only to motivate the use of the Gibbs/Shannon/Jaynes entropy as a measure of the uniformity (or information content) of any physical distribution, which it might seem appropriate to maximize (or minimize). However, the real justification comes from arguments such as those of Chapter 2 where certain assumptions on the structure of images lead naturally to the *configurational* entropy reflecting the prior probability of an image. The message is that the entropies of positive images and probability distributions are rather different beasts.

Secondly, the majority of applications of MaxEnt in this book use the constraint $\chi^2 = N$ to incorporate the data. This is a somewhat *ad hoc* technique, justified by the fact that, on average, every data point is one standard deviation away from its true value. Thus it is a 'long run' or *frequentist* rule. In practice what this constraint does is to set the relative weight given to the entropy and to the data (via χ^2) in determining the resultant image: it balances uniformity against (possibly spurious) structure. An alternative and Bayesian way to proceed is to enlarge the *hypothesis space* to include the relative weighting of these terms among the parameters to be estimated. This is the role played by α in Chapter 2. It will in general lead to a χ^2 not equal to N.

Macroirreversibility and microreversibility reconciled describes how the conceptual difficulties of the second law of thermodynamics disappear when the methods of statistical physics are recognized as instances of reasoning from incomplete information. The second law describes the loss of information which results when a system is perturbed and note is taken only of its final equilibrium state, not of its intermediate dynamical evolution. The thermodynamic entropy should be identified with the maximized Gibbs/Shannon/Jaynes entropy. The ideas are then generalized to set up a formalism for non-equilibrium processes. In fact this chapter is a concentrated crash-course in Bayesian techniques and its arguments are quite general. In three appendices, further side-issues are addressed. Again a certain mathematical sophistication is assumed of the reader.

In *Some misconceptions about entropy*, Dr Garrett's arguments of the previous chapter are amplified and illustrated. The Boltzmann H function, an early candidate for the statistical representation of thermodynamic entropy, is shown to be wanting, failing to describe other than non-interacting systems. Dr Gull demonstrates in a new way how Brownian motion, an irreversible process with which orthodox statistical physics has had some difficulty, can be accounted for. This can be done by the introduction of a maximum entropy probability distribution to describe the space–time trajectories of particles, with constraints coming from the dynamics of the process. This chapter closely follows the original talk and some of the derivations are rather compressed.

The final chapter, *The X-ray crystallographic phase problem*, presents a brief review of current ideas on how to tackle the extremely demanding inverse problem that arises in the analysis of X-ray diffraction patterns. There emerges an alternative derivation of the entropic prior which is couched to some extent in the language of orthodox statistics. It is useful to have different derivations of central results and Dr Bricogne's arguments, using as they do the saddlepoint approximation, recall the famous 'justification' of the maximum entropy distribution which goes under the name of the Darwin–Fowler method.

Section headings

1 Of maps and monkeys: an introduction to the maximum entropy method

G. J. Daniell

2 Fundamentals of MaxEnt in data analysis

J. Skilling

3 Maximum entropy and nuclear magnetic resonance

P. J. Hore

4 Enhanced information recovery in spectroscopy using the maximum entropy method

S. Davies, K. J. Packer, A. Baruya and A. I. Grant

5 Maximum entropy and plasma physics

G. A. Cottrell

8 The X-ray crystallographic phase problem

G. Bricogne

About the authors

Babul Baruya

BP Research Centre
Chertsey Road
Sunbury-on-Thames
Middlesex, TW16 7LN

A. Baruya is a senior statistician at BP Sunbury. He specializes in the development and application of advanced statistical and data processing techniques.

Gérard Bricogne

MRC Laboratory of Molecular Biology
Hills Road
Cambridge, CB2 2QH
and
LURE
Bâtiment 209d
91405 Orsay
France

Dr G. Bricogne was born and educated in France, where he graduated in mathematics and chemistry. He holds a Ph.D. from the University of Cambridge, for research done at the MRC Laboratory of Molecular Biology on phase determination methods based on the exploitation of non-crystallographic symmetry. The computational aspects of this work led to the first determinations of virus structures to atomic resolution. His subsequent research has been devoted to extending 'direct' methods of X-ray crystal structure analysis, well known for small molecules, to macromolecules. His interests lie in the methods of molecular structure determination and representation, and especially in the underlying mathematics.

Dr Bricogne has held positions as a research fellow of Trinity College,

Cambridge (1975–81), as an assistant professor of biochemistry at the College of Physicians and Surgeons of Columbia University, New York (1981–83), and as a director of research at the French Synchrotron Radiation Facility (LURE) in Orsay (1983–present). Since 1988 he has been working as a visiting scientist at the MRC Laboratory of Molecular Biology and at Trinity College, Cambridge.

Geoff Cottrell

JET Joint Undertaking
Abingdon
Oxfordshire, OX14 3EA

G. A. Cottrell is principal scientific officer in the physics group (radio frequency heating) at the Joint European Torus (JET) experiment based at the UKAEA's Culham Laboratory in Oxfordshire, UK. After reading physics at Sussex University, he received his Ph.D. in radio astronomy at the University of Cambridge in 1977. He spent two years in postdoctoral research in low temperature solid state physics at UMIST, before transferring to Culham Laboratory in 1979 to work on controlled thermonuclear fusion. He joined the JET project full-time in 1985. Current areas of study include the interpretation of plasma physics experiments on energy confinement and heating as well as the development of a physical model capable of both explaining present experimental results and predicting future tokamak performance. He is currently a research fellow at Wolfson College, Oxford.

Geoff Daniell

Department of Physics
The University
Southampton, SO9 5NH

G. J. Daniell read natural sciences at Downing College, Cambridge and went on to gain a Ph.D. at the Cavendish Laboratory, on the theory of a radio aerial immersed in the ionosphere. He is currently a senior lecturer in the Department of Physics at the University of Southampton. His research interests include data and signal processing techniques, maximum entropy and the foundations of probability theory and computational techniques in physics with especial emphasis on problems in quantum field theory and statistical mechanics.

Simon Davies

BP Research Centre
Chertsey Road
Sunbury-on-Thames
Middlesex, TW16 7LN

S. Davies joined BP Research in 1988 after completing his doctorate in nuclear magnetic resonance (NMR) spectroscopy at Oxford University. His interests include frequency-selective excitation, the use of relaxation data to probe the internal structure of porous media, NMR imaging and inverse problems.

Anton Garrett

Department of Physics and Astronomy
University of Glasgow
Glasgow, G12 8QQ

A. J. M. Garrett took his first degree in physics at the University of Cambridge. He remained there and gained his Ph.D. in kinetic theory in 1984. It was while he was a research fellow at Magdalene College that his conversion to the Bayesian outlook occurred, through his contact with Steve Gull. From 1985–88, he held a research fellowship at the University of Sydney and currently he is a Royal Society of Edinburgh research fellow at the University of Glasgow.

His interests are wide-ranging with most of his recent work centred on quantum philosophy and the foundations and applications of probability theory. On the latter, he is at present writing a book, to be called *'Inference and Inferential Physics'*.

He is a member of UK Skeptics and is a keen debunker of pseudo-science.

Andrew Grant

BP Research Centre
Chertsey Road
Sunbury-on-Thames
Middlesex, TW16 7LN

A. Grant joined BP Research in 1984 after completing his doctorate in time-resolved, laser flash photolysis electron spin resonance at Oxford

University. His interests include the application of appropriate laser and optical spectroscopies to industrial problems, and the recovery of enhanced information from spectroscopic data, recorded under industrially relevant conditions.

Steve Gull

Cavendish Laboratory
Madingley Road
Cambridge, CB3 0HE

S. F. Gull took his first degree in theoretical physics at the University of Cambridge. His Ph.D. was earned in the radio astronomy group at the Cavendish Laboratory there. He has remained a member of St John's College and is now a university lecturer in physics.

Dr Gull has done work in the dynamics of radio galaxies and supernovae, the astronomy of the cosmic background and numerical hydrodynamics and plasma physics. He was one of the leading figures in the establishment of the maximum entropy method applied to image enhancement and now works on general maximum entropy data processing, the foundations of probability theory and inverse problems of all types. Currently he is working extensively on the physical applications of Clifford algebras.

With John Skilling, he co-founded Maximum Entropy Data Consultants Ltd.

Peter Hore

Physical Chemistry Laboratory
South Parks Road
Oxford, OX1 3QZ

P. J. Hore is a university lecturer in Physical Chemistry and fellow of Corpus Christi College at the University of Oxford. Having got both his degrees from Oxford, he spent two years as a Royal Society European programme research fellow with Professor R. Kaptein at the University of Groningen. He returned from the Netherlands to a junior research fellowship at St John's College, Oxford to work with Dr (now Professor) R. Freeman, and took up his present position in 1983. His research interests include energy conversion in photosynthetic reaction centres, spin effects in chemical reactions, protein structure and folding, and the development of new methods in nuclear magnetic resonance spectroscopy.

Ken Packer

BP Research Centre
Chertsey Road
Sunbury-on-Thames
Middlesex, TW16 7LN

Professor K. Packer is a chief research associate with BP Research at their Sunbury-on-Thames (UK) research centre. He manages the spectroscopy, microscopy and structural crystallography activities within the Analytical Support and Research Division. Prior to joining BP Research in 1984, he held a personal chair in the School of Chemical Sciences, University of East Anglia, Norwich. His research interests centre on nuclear magnetic resonance and its application to a wide range of physico-chemical areas.

John Skilling

Department of Applied Mathematics and Theoretical Physics
Silver Street
Cambridge, CB3 9EW

J. Skilling was an undergraduate at Cambridge, obtaining his first degree in physics in 1965, following which he joined the radio astronomy group in the Cavendish Laboratory for his doctoral work on plasma instabilities in astrophysics. He spent two years abroad at Princeton University (1969–71) working at the Plasma Physics Laboratory, and then returned to Cambridge to join the teaching staff of the Department of Applied Mathematics and Theoretical Physics, with cosmic ray physics as his major research interest.

In 1977, his research career changed abruptly through the seminal work on maximum entropy image reconstruction by Gull and Daniell. Since then, he has concentrated on this field, contributing to both theory and practice. In 1980, he founded Maximum Entropy Data Consultants Ltd with Dr Gull, to assist in the development of a wide variety of practical industrial and academic applications of maximum entropy techniques.

1

Of maps and monkeys: an introduction to the maximum entropy method

G. J. Daniell

Abstract

We present a tutorial introduction explaining why an approach like maximum entropy is essential in a wide range of problems in image and data processing. The simple 'monkey argument' and Bayes' theorem are used to justify the use of entropy and a typical application is discussed which illustrates the philosophy of the maximum entropy approach.

1.1 Introduction

In scientific investigations of the physical world, we design apparatus, make measurements and process these measurements to arrive at conclusions. In a good experiment each of these three aspects is considered critically; I am concerned here with the third: the data processing.

We usually take lots of measurements. The readings obtained from the apparatus are not usually perfectly reproducible and some averaging operation, at the very least, is desirable to convert them into a single value and an estimate of its uncertainty. In most experiments the conclusion is not just a single number and I show that there exists a class of difficult problems where the conclusions potentially outnumber the data. I describe the maximum entropy method for attacking these problems, giving some examples and showing some results. There is a subsidiary theme. I am advocating Bayesian methods of data processing and I show that both a familiar operation such as calculating the average of a set of numbers and also the maximum entropy method are examples of a unified Bayesian approach.

What I shall say applies to many different problems in physical science, so I will start by talking about data processing in a very general way.

1.2 The general philosophy of data processing

Data processing is the conversion of data to conclusions about the physical world. I can represent the operation as a diagram (Fig. 1.1).

An elementary point concerning the data is that they comprise a finite set of numbers. With computer-controlled experiments it is obvious that the set of numbers is finite. With a chart record, we can always digitize the record with several points in the width of a pen line. The question of whether such readings are truly independent must not be forgotten, but we can indisputably convert all that is worth knowing about a chart record to a finite set of numbers.

I emphasize that, by the data, I mean the exact numbers read from scales on the instruments. I do not mean those numbers 'corrected' or 'calibrated', or the numbers with the resolution of the instrument removed. Traditionally, scientists start with readings and carry out a series of operations of 'correction' or 'calibration', by which these numbers are transmuted, step by step, into conclusions. I reject this approach as fundamentally wrong. If I perform a calibration experiment, the readings are part of the data set of a larger experiment and the calibration curve is, incidentally, part of the conclusions. Every reading of every instrument belongs in the data, on the LHS of Fig. 1.1.

One useful way of classifying the conclusions is to count how many numbers are needed to describe them. I can then consider various cases:

1. A logical conclusion: A's theory is much less likely to be correct than B's;
2. One number: The velocity of light is ...;
3. A few parameters: The position, width and intensity of a spectral line, of predictable shape given by an indisputable theory, are ...;
4. A smooth, but otherwise arbitrary, curve: An isotherm of an imperfect gas is a good example.

   There are a lot of smooth curves—an arbitrary smooth curve requires an infinite number of parameters to specify it—but we can get a very good approximation to an arbitrary smooth curve with a finite number of parameters. Let us say that our conclusions in this type of experiment are a small infinity of numbers.
5. A 'rough' curve: this is best illustrated by some examples.

   Consider an astronomical object and suppose we want a map of the brightness distribution across it. Every time we build a bigger telescope we see more detail, and we can imagine continuing to build bigger telescopes, until we can see every atom in the universe. The brightness distribution of such an astronomical object need not be smooth: it can vary arbitrarily from point to point. Because of the

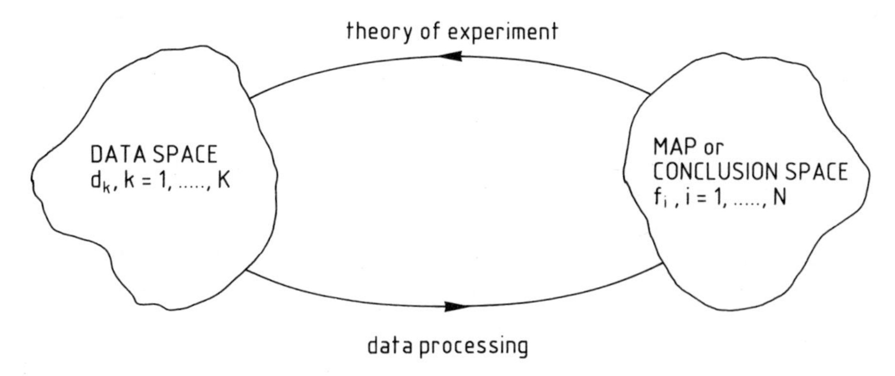

Fig. 1.1. The theory of the apparatus enables us to predict the data. Data processing is going from the data to the conclusions.

astronomical application, I will use the word *map* to describe such 'rough' curves.

Most problems in spectroscopy also fall into this class. Every time we build a better spectrometer, we can resolve closer lines and finer splittings and improvements in signal-to-noise ratio reveal weaker lines. The exact spectrum always remains unattainable. A 'rough' curve clearly requires an infinity of parameters to describe it and cannot be approximated by a small number of them.

I can summarize data processing as determining N numbers from K observations. K is finite. N may be small or it may be infinite. There are obviously four cases:

A: $N < K$. The data, regarded as exact values, are usually mutually inconsistent,

B: $N > K$. The problem is insoluble: there are many conclusions consistent with the same data,

C: $N = K$. There is a unique solution according to the mathematicians,

D: $N = \infty$. Very insoluble!

I will examine data processing strategies for these different cases.

1.3 Data processing strategies

1.3.1 A: The case $N < K$: inconsistent data and experimental errors

Consider the following idealized problem. I have measured the length of a piece of string three times, and I have three different numbers. I want one number for the length. This problem was 'solved' by Gauss by the invention of the concept of 'experimental error'. We are all so familiar with this treatment of data, that we fail to see what an enormous conceptual step is involved. I no longer consider just my one piece of string, but an ensemble of pieces, and I pretend that my three measurements are the lengths of three pieces of string in the ensemble. I also change the question; I no longer want the length of my piece of string, but the mean length of the pieces in the ensemble. No wonder students find statistics confusing!

The Bayesian approach to this problem is much more convincing, and I will make a diversion to explain it. For the present purposes Bayes' theorem can be stated as

$$\begin{aligned}&\text{Prob(conclusion} \,|\, \text{new data)} \\ &\quad \propto \text{Prob(conclusion} \,|\, \text{old data)} \times \text{Prob(new data} \,|\, \text{conclusion).}\end{aligned} \tag{1.1}$$

The notation Prob$(A|B)$ denotes the probability that A is true, given that B is true. This theorem tells us how to change our conclusions when we get more data. I can apply this to the three measurements of the string. Let the conclusion be the proposition that the actual length of the string is between L and $L + \mathrm{d}L$. The new data are my three measurements of the length x_1, x_2 and x_3. Suppose that the old data is my estimate, by eye, that its length is about 1.5 m and certainly between 1 and 2 m so that Prob(conclusions | old data) looks something like Fig. 1.2a. Suppose also that I know my measurements have a Gaussian distribution, with standard deviation σ. Then if the true length is L, the probability of measuring x is

$$\frac{1}{\sqrt{2\pi}\,\sigma} \exp\left(-\frac{(x-L)^2}{2\sigma^2}\right)$$

and the second term on the RHS of (1.1), called the likelihood, is

$$\text{Prob(new data} \,|\, \text{conclusion)} \propto \prod_{i=1}^{3} \exp\left(-\frac{(x_i - L)^2}{2\sigma^2}\right).$$

If I plot this as a function of L, I get a curve like Fig. 1.2b. Bayes' theorem says that I should multiply the probability distributions of Fig. 1.2a,b. It is clear that because the peak in Fig. 1.2b is much narrower than the peak in Fig. 1.2a, the exact shape of the curve in (a) is unimportant. Whatever reasonable assumption I use as old data, I get the result that

$$\begin{aligned}&\text{Prob(string is between } L \text{ and } L + \mathrm{d}L \,|\, 3 \text{ measurements)} \\ &\qquad \propto \exp\left(-\sum_i \frac{(x_i - L)^2}{2\sigma^2}\right).\end{aligned}$$

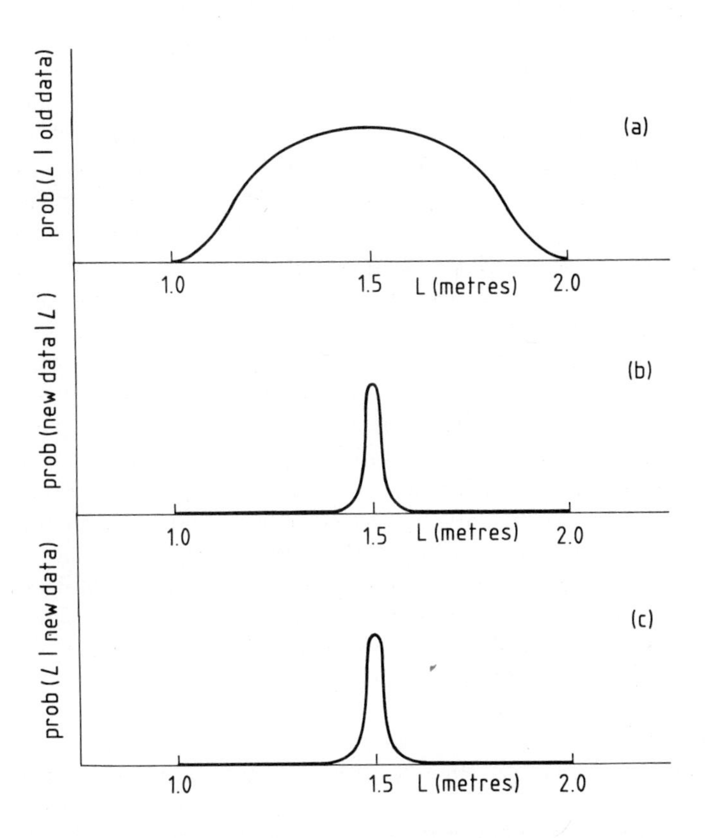

Fig. 1.2. (*a*) The prior probability distribution for the length of the string before any measurements. (*b*) The likelihood of getting different lengths for the string, denoted by Prob(new data | conclusion) in the text. (*c*) The posterior probability distribution for the length of the string, Prob(conclusion | new data): this is given by Bayes' theorem as the product of the distributions in *a* and *b*.

This is as shown in Fig. 1.2*c*. This graph summarizes all that I know, after the experiment, about the length of the piece of string. Notice that only my string is mentioned, not an infinite number of other bits of string. If I want just one number to describe its length, then I can choose to quote the value of L at the maximum of the probability distribution, or the mean of the distribution. Instead of the whole curve, I can conveniently use the width as a measure of the range in which the length of the string might reasonably lie. In this example, the maximum and the mean both occur at the same value of L, namely $L = \frac{1}{3}(x_1 + x_2 + x_3)$. The fact that this is the average of the measured values is a consequence of the Gaussian distribution of the errors. If this distribution is not Gaussian, the maximum and the mean may not coincide; neither may equal the average of the measurements; and it may not even be appropriate to quote one single value of L to summarize the result of the experiment. This example of a Bayesian calculation has assumed that the standard deviation of the measurements of the string length is known beforehand. It is sufficient here to assure the reader that the Bayesian approach can easily be extended to the more realistic case where it is not (see, for example, the book by Jeffreys (1961, pp. 137–40)).

The Bayesian approach is so simple, why does everyone not use Bayesian methods? The complications occur when the data are bad. With good data the first term on the RHS of (1.1), called the prior probability, is unimportant. If the spread of the measurements approaches a metre, then it is not good enough to use as the 'old data' my visual judgement that the length is about 1.5 m. In a general statistical problem, we do not know how to calculate the prior probability when we have no data at all and because of this, many statisticians reject Bayesian methods. But for most simple problems, we *can* assign a sensible prior and it is worth stressing that the prior only matters when the data are so poor that, after the experiment, we are still appreciably influenced by the prejudices we had beforehand. Because of their simplicity, I believe that Bayesian methods are more useful to physical scientists than conventional statistics. I now return to the general data processing problem.

1.3.2 B: The case $N > K$: insufficient data and inverse theory

Professional statisticians are remarkably silent about what to do in this case, so let us examine what people do in practice. Two common approaches are:

1. Model fitting.

   The idea is to assume that the 'rough curve' being sought is actually smooth and has a simple functional form; we might, for example, assume that a spectrum contains only two spectral lines and that their shapes are Lorentzian.

   The problem is then changed to one of parameter fitting, with $N < K$. This is a risky procedure: we get an answer whether the assumption is true or false. Either we believe the model, in which case the problem had $N < K$ from the start, or we don't know.

   If we don't know whether the model is correct then all we have done is to construct one of the infinity of conclusions that fits the data, and we don't even know whether the one we have found is typical.

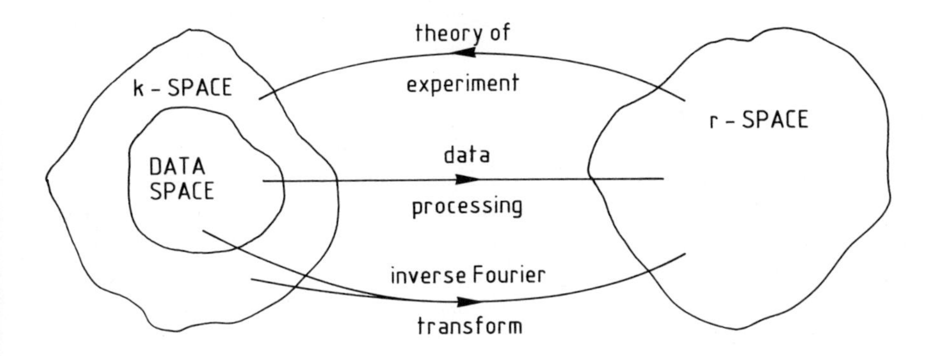

Fig. 1.3. An experiment involving a Fourier transform. The relation between data processing and the inverse transform.

Construction of a solution is easy; the difficult part is discovering what features of that solution follow inexorably from the data and which ones are a result of the model. Model fitting is easy, common, and dangerous.

2. Invent sufficient extra data to get $N = K$ and use the mathematician's solution.

   I am deliberately using this inflammatory language; I want people to understand that this common approach is almost as bad as changing experimental values that don't look right.

   The invention of the data is usually hidden. A very common example is the Fourier transform and I will discuss this in detail, since it is also used for the numerical illustration below. Suppose we want a function in r-space and the data is obtained, by a scattering experiment, for example, in k-space, where there is a Fourier transform relation between k- and r-space. If we had measurements for all values of k then the inverse Fourier transform would immediately solve our problem. However, the data are usually imperfect because:

   1. There is a maximum k for which we have measurements.
   2. The k-space is sampled and there may be gaps.
   3. Detector resolution results in data which may be averages over a range of k.

   The serious limitation is usually the first, although the approach that I shall describe later deals with the other limitations at the same time. A diagram illustrating the problem is Fig. 1.3. The inverse Fourier transform takes all of k-space into r-space, whereas data space is only a part of k-space. A frequent assumption is that the Fourier components that have not been measured are zero in amplitude. The only justification for picking the value zero rather than any other is that a large value would clearly introduce large oscillations in r space. The consequences of choosing zero are, however, well known; the resulting functions in r-space have oscillations caused by the discontinuity at the maximum measured k. These are frequently so pronounced that the solution in r-space has to be smoothed to reduce them. This approach to Fourier transform problems is so bad that one is forced to reduce the resolution, that is, throw away good data, to make the result even acceptable. After this the solution will still contain oscillations that may resemble the features we are trying to observe. All this is unnecessary if we use a good method of data processing, such as maximum entropy.

1.3.3 C: The case $N = K$ and unstable problems

If $N = K$, a mathematician may conclude that there exists a unique solution to our data processing problem and he can often produce a formula which appears relevant to the inverse problem. There is sometimes a serious limitation with such analytic solutions: they can be unstable. A common example is deconvolution. Another well-known one is the inversion of the Laplace transform. An unstable problem occurs when a lot of data values contain almost the same information about our conclusions. If, given two measurements, I can predict to a good approximation, the value of a third measurement, then the effective number of measurements is less than the actual number. What is on the surface an $N = K$ problem becomes an $N > K$ problem, when the experimental errors are considered. Using the mathematician's solution in an unstable problem results in enormous amplification of experimental errors, or the inversion of almost singular matrices, and frequently produces a complaint to the computer advisory service that a library program does not work. There is a large literature on unstable problems. The point here is that it is better to classify them along with the $N > K$ problems and the same methods of solution will then apply.

1.3.4 D: The case $N = \infty$

It is clear that the problem of determining a 'rough' curve is simply an extreme case of the $N > K$ problem.

To conclude this section, Table 1.1 shows some examples of map spaces, data spaces and the operations relating them.

Table 1.1. Map spaces, data spaces and the operations relating them.

Field	Map space	Data space	Operator
Optical astronomy	True sky	Blurred sky	Convolution
Radio astronomy	Radio sky	Fringe visibility	Fourier transform
Fourier spectroscopy	Spectrum	Fringe visibility	Fourier transform
Magnetic resonance	Spectrum	Free induction decay	FT + damping
X-ray tomography	X-ray absorption	Line integrated density	Radon transform
Small angle neutron scattering	Size distribution	Scattering func. $S(q)$	Similar to FT
Small angle neutron scattering	Size distribution	Poorly resolved $S(q)$	Approx. FT + convolution

1.4 The Bayesian approach to the $N > K$ case: maximum entropy

Bayes' theorem gives us

$$\begin{aligned}&\text{Prob}(\text{map} \mid \text{new data}) \\ &\qquad \propto \text{Prob}(\text{map} \mid \text{old data}) \times \text{Prob}(\text{new data} \mid \text{map}).\end{aligned}$$

In our previous Bayesian example ($N < K$), the conclusion followed essentially from the second factor; only one length for the string gave a high probability of producing the data. The term Prob(map | old data) was essentially irrelevant and it did not matter that a rough and ready visual estimate was used for it. Now with $N > K$, the roles of the terms are reversed. The data do not sufficiently constrain the map: there are lots of maps for which Prob(new data | map) is large. We must decide which of these maps we prefer by assigning Prob(map | old data). In most cases, the maps that are consistent with the data differ in the fine details that the data are not adequate to resolve. In these cases, we want to choose the prior probability to select smooth maps preferentially. Maximum entropy is one model for the prior that does this. Note that I am adding assumptions about smoothness in *map* space, in contrast to the argument for assuming zero-valued Fourier components, which had the same aim, but was applied in *data* space.

There is another difference between the two applications of Bayes' theorem. In my string measurement, it was possible to draw the probability distribution Prob(length | data) (Fig. 1.2*c*). In the case where map space has an infinite number of dimensions, I cannot plot this probability. The only realistic way of presenting the results of the experiment is as a single map, and this can be obtained by maximizing the probability.

Entropy is a concept in probability theory and the maximum entropy method is applicable when we are determining a function that can be regarded as a probability distribution. Examples of this are the number of particles as a function of energy (that is, most types of spectroscopy), electron densities, distributions of particle sizes, optical, radio or X-ray images.

The entropy method can be justified in several ways. Skilling (1989) (see also the next chapter) has given some axioms which lead to it. I will present the *monkey* argument (Jaynes 1986). Imagine we are determining a function $f(x)$. Divide x-space into cells of width Δx, which is much less than the conceivable resolution of the experiment, so that no approximation is involved. Let the intensity of $f(x)$ in the cell number i be f_i and suppose that the intensity is quantized so that $f_i = n_i\delta$. The quantum δ is so small that again this is not an approximation and the numbers n_i are all large. I can represent the function $f(x)$ by a set of integers $n_1, n_2, \ldots$.

Now imagine the traditional team of monkeys, throwing balls into boxes, which are the cells of x. They do this completely at random, with no regard for any experimental data. The probability that they produce $n_1, n_2, \ldots$ is simply the combinatorial expression

$$2^{-M}\frac{M!}{n_1!\,n_2!\,\cdots},$$

where M is the total number of balls. Since the n_i are large, I can use Stirling's approximation and get

$$\mathrm{Prob}(f_1, f_2, \ldots) \propto \exp(\alpha S), \tag{1.2}$$

where

$$S = -\sum_i f_i \log\frac{f_i}{b}$$

and α and b are constants depending on M and δ. The quantity S is called the entropy of the function $f(x)$, and we use the expression (1.2) for Prob(map | old data).

From a map $f(x)$, we can predict what the observations should be and get the residuals r_i, defined as the difference between the ith datum and the value for this datum predicted from the map $(f_1, f_2, \ldots)$. This calculation involves the solution of the forward problem, namely going from map space to data space. If the r_i are not zero, it is because of experimental errors in the data. If these have a Gaussian distribution, Prob(data | map) is proportional to

$$\prod_i \exp\left(-\frac{r_i^2}{2\sigma^2}\right) = \exp(-\tfrac{1}{2}\chi^2).$$

So Bayes' theorem gives

$$\mathrm{Prob}(\mathrm{map}\,|\,\mathrm{new\ data}) \propto \exp\big(\alpha S - \tfrac{1}{2}\chi^2\big).$$

Maximizing this is the same as maximizing $\alpha S - \frac{1}{2}\chi^2$, which is to say, the maximum entropy map is obtained by maximizing

$$-\alpha \sum_i f_i \log\left(\frac{f_i}{b}\right) - \tfrac{1}{2}\chi^2(f_1, f_2, \ldots). \tag{1.3}$$

The constant α is arbitrary and is frequently chosen to make χ^2 equal to the number of observations K, since this is the expected value for a χ^2 distribution. This is certainly not the definitive rule and alternative possibilities are discussed by Gull (1989).

1.5 Maximum entropy in practice

We do not need real monkeys, or even a computer simulation of monkeys; we have calculated how often they produce each map. This leads to a problem in numerical analysis of finding the maximum of a function of a large number of variables. Skilling and Bryan (1984) have developed an algorithm that works well for most problems and has been used for maps with a million points, implying a maximization in a million variables! For linear problems, that is those where the data are related to the map by a linear operator, it is possible to show that there is only one maximum, so the maximum entropy solution is unique. The computation is relatively slow as it is iterative and requires the calculation of the forward problem to predict the data. The time required is approximately 100 times that for the solution to the forward problem. For a one-dimensional Fourier transform, only a few seconds of CPU time are required. Collecting the data usually takes a lot longer than this, and is a lot more expensive. I am tired of hearing people reject maximum entropy because it is slow.

1.6 Understanding maximum entropy solutions

We are selecting one map from all possible maps on the basis of its entropy. The selection is made on a criterion in map space, so we can immediately interpret the effects of our choice. Contrast this with making some Fourier coefficients zero, where the constraint is imposed in data space and we cannot easily interpret the consequences. The following general features of the maximum entropy map may be anticipated.

1. We automatically get a positive map: the monkeys do not throw negative balls, so they generate only positive maps.
2. The maximum entropy solution is the most uniform one consistent with the data and its errors.

This can be seen in an example as follows. Suppose we are sharpening a blurred spectrum by removing instrumental resolution. Consider three adjacent cells; the monkeys will throw different numbers of balls into these boxes in different trials. Suppose that the spectrometer cannot resolve these cells at all, so that the observed spectrum depends only on the total flux in the three cells. The constraint of the data, through χ^2, fixes the total flux, but the distribution of the flux between the cells is determined only by the contribution of these cells to the entropy. The maximum entropy spectrum has equal fluxes in all three cells, since the monkeys produce this configuration more frequently than others.

3. Any departure from uniformity that is apparent must be essential for the map to fit the data.

   Because of the pressure towards uniformity, if a departure from uniformity appears, then it must be due to something in the data, provided the experiment has been correctly understood and realistic errors used. A maximum entropy map should therefore not contain artefacts arising from the data processing.

4. In an unstable problem, where the data do not determine the solution properly, then a reproducible uniform solution results without noise amplification. Simple modifications to the method permit any 'default' answer to be specified.

5. Because of the pressure towards uniformity, peak values are too low and valleys are too high.

   A maximum entropy solution is therefore biased in the statistical sense. Because this is frequently said to be a defect of maximum entropy, it is worth discussing the implications in more detail. When analysing real data, we will never detect bias, because we do not know the right answer, so let us envisage a test using simulated data. Suppose the experiment is to determine the value of the intensity of a spectral line. We generate several synthetic data sets using a 'true' intensity, include some simulated experimental errors, analyse these data sets using maximum entropy, and read off the intensity of the line in each of the reconstructed spectra. Because of the experimental errors, these intensities will not all equal our 'true' value, but will have some distribution. The bias inherent in the maximum entropy method implies that the mean of this distribution will be lower than the 'true' value. However, there is no fundamental reason for considering the mean at this point. Physicists are used to considering arithmetic means, but they are also used to dealing with symmetric distributions, where all sensible measures of location are equal. The point is that there is no absolute objection to bias in an estimation procedure.

This example is also a misuse of maximum entropy. If all that is required is the intensity of a single spectral line, then curve fitting is the best way of getting this. If we want a general-purpose picture of the spectrum, then bias is less important and maximum entropy is a very good approach.

1.7 An example

As an illustration, I consider the following simple Fourier transform problem. Suppose I want to determine a function $f(x)$ with $x \in [0, 1]$ and that the data are the eight numbers d_k, given by

$$d_k = \int_0^1 f(x) \cos(\pi k x)\, dx \qquad \text{with} \qquad k = 0, \ldots, 7. \tag{1.4}$$

The values I have analysed are shown in the following table:

k	d_k
0	120.00
1	34.20
2	−76.12
3	−77.60
4	−23.41
5	64.25
6	105.83
7	−3.92

The maximum entropy solution for $f(x)$ is a smooth function. For the purposes of numerical computation, I represent it by a set of N sample values, and I can approximate the integral in (1.4) by the sum

$$d_k = \frac{1}{N} \sum_{n=0}^{N-1} f_n \cos(\pi k n / N), \tag{1.5}$$

and this sum can be conveniently evaluated by the FFT algorithm. The number of points N must be chosen to be so large that replacing the integral by the sum produces negligible error.

The maximum entropy solution depends on the value of the parameter b in (1.3) and on the errors associated with the data values. I have taken b to be very small and then the results hardly depend on it and I have also used the $\chi^2 = K$ rule to fix α. Fig. 1.4*a* shows the maximum entropy solution with $N = 64$ and errors of ± 5 on each of the d_k. The curve is seen to be adequately sampled and this is confirmed in Fig. 1.4*b*

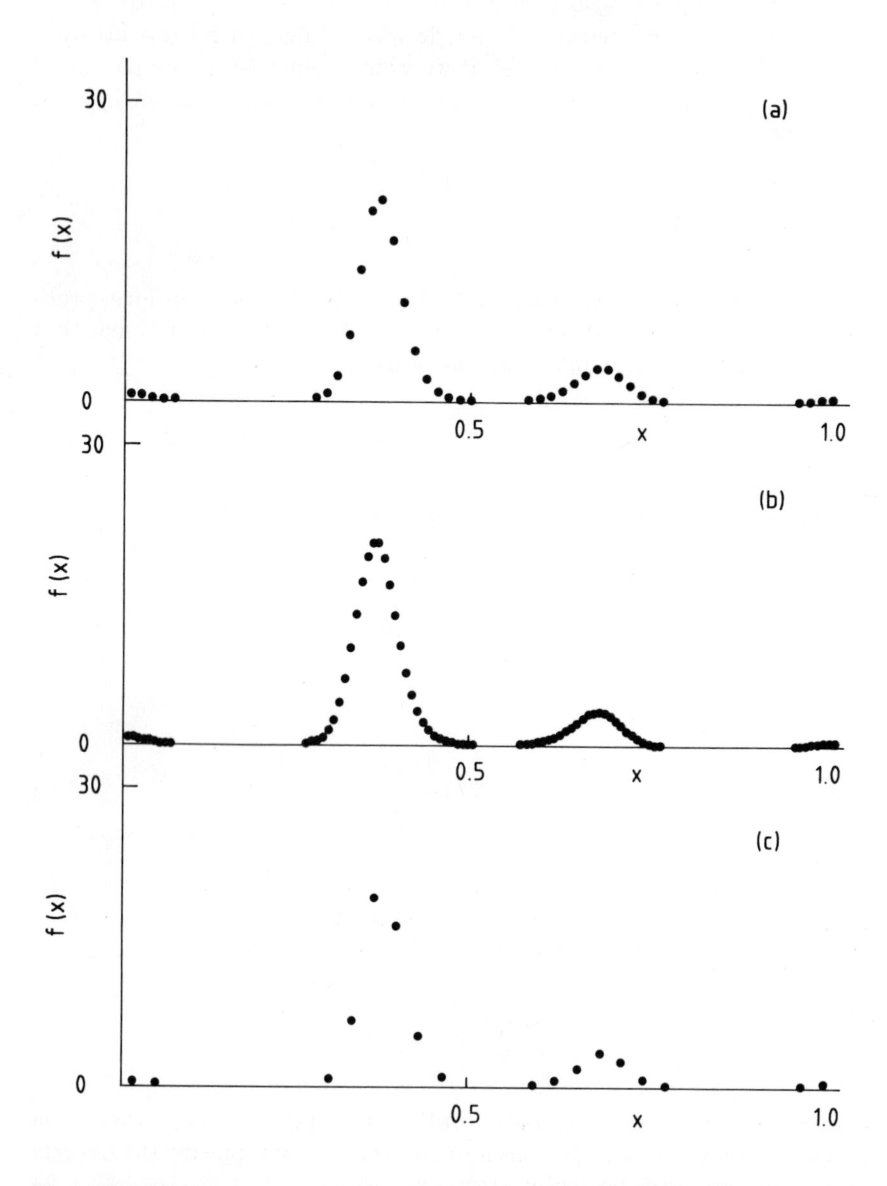

Fig. 1.4. Maximum entropy solutions to the inverse problem of (1.4), with a noise level of ± 5 on each data point. (*a*) The function $f(x)$ sampled by 64 points. The structure is adequately sampled. (*b*) The maximum entropy solution to the same problem computed with 128 samples. No extra resolution is produced: the extra points are simply interpolated on a smooth curve. (*c*) The same problem, but with only 32 points, showing serious undersampling of the structure present.

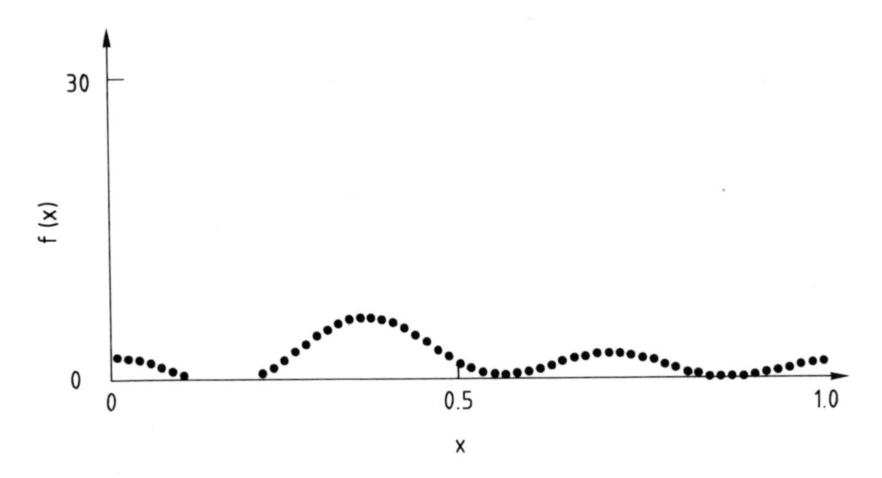

Fig. 1.5. The truncated inverse Fourier series solution to the same problem of (1.4). The parts where the solution is negative are suppressed.

which is calculated with $N = 128$. This leads to no change of shape but simply the interpolation of extra samples in the solution. For comparison, Fig. 1.4c was computed with $N = 32$ and shows serious undersampling of the solution.

It follows from (1.4) that

$$f(x) = 2\sum_{k=0}^{\infty} d_k \cos(\pi k x),$$

and if we knew all of the d_k, then this equation would be useful. Substituting the observed values of d_0 to d_7 into this and taking $d_k = 0$ for $k > 7$ gives the curve of Fig. 1.5; the parts where the curve goes negative are not shown. Either of Figs 1.4a or 1.5 could be the solution for $f(x)$ that we require; any preference between these cannot be made on the basis of the data, but possibly because Fig. 1.4a has the properties 1 to 5 listed above.

The difference between this approach involving the invention of data, and maximum entropy is illustrated by Fig. 1.6, where the values of the RHS of (1.4), for $k = 0$ to 32, corresponding to the maximum entropy solution have been plotted (crosses) together with the eight experimental values of the d_k from the table and their errors. The first eight values are constrained to get good statistical agreement with the experimental values and the remaining ones have been freely adjusted by the algorithm to maximize the entropy of the solution for $f(x)$.

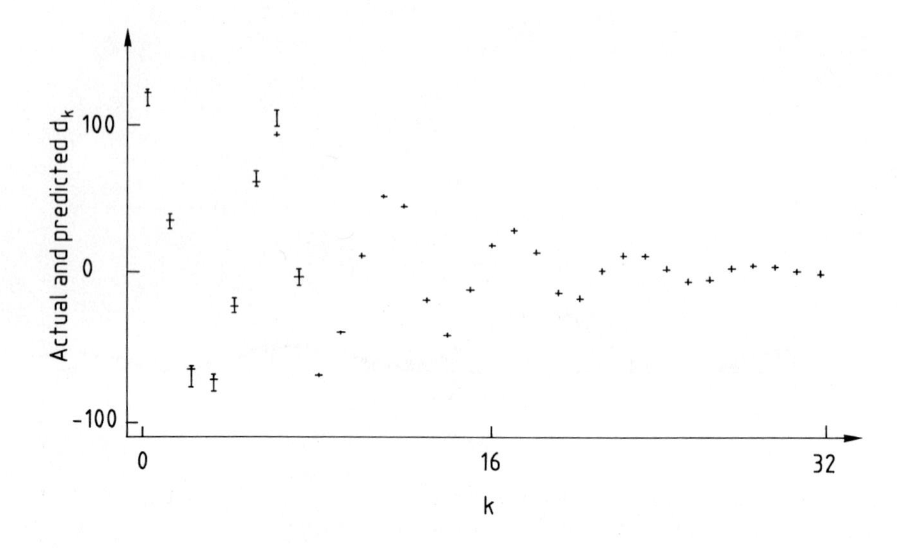

Fig. 1.6. The agreement between the data values predicted from the maximum entropy solution and the observed values. Values of d_k for $k = 0$ to 32 are shown but observed values correspond only to d_0 to d_7.

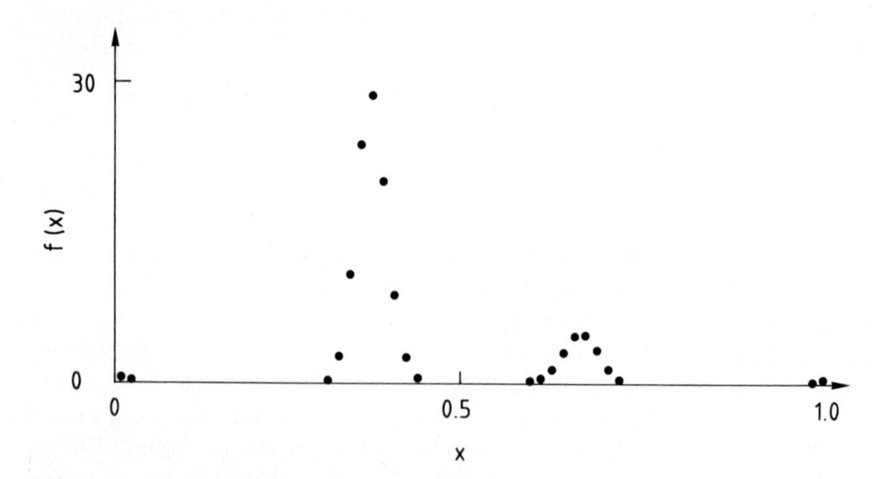

Fig. 1.7. The dependence of the maximum entropy solution on the errors in the data. The errors associated with each data point are ± 2. Contrast this with Fig. 1.4 where the errors are ± 5. The stronger constraint imposed by the data has produced a narrower peak.

One occasionally meets an approach equivalent to taking $N = 8$ in (1.5) and using the inverse discrete Fourier transform. This is completely wrong. Equation (1.5) represents a different problem, which is equivalent to that of (1.4)—the one we are trying to solve—only if N is large. I am looking for a function $f(x)$; sampling is just a convenient way of representing it, and we have seen that even 32 samples are insufficient.

The dependence of the maximum entropy solution on the errors in the data is illustrated by Fig. 1.7 which shows the $N = 64$ solution when the error associated with each of the data values in the table has been reduced to ± 2. The stronger constraint imposed by the data forces the solution to be less uniform and to have lower entropy. If these graphs represented spectra, one would say that the resolution had been increased.

1.8 Return to the general philosophy of data processing

Data processing is getting from data space to map space. The theory of the experiment can be regarded as an operator from map space to data space. Applying this operator—the forward problem—is always possible provided we can analyse the experiment.

We assume we know how the apparatus works; we have a theory that accounts for all the steps in the process of getting from the physical quantity we are interested in to the data we collect. Consider, for instance, a scattering experiment. Perhaps we are not using a completely monochromatic source. There may be scattering processes that produce a background. Some of the data may be ruined by other physical effects, a Bragg reflection, for example. Perhaps some data cannot be collected because of a support in the apparatus. Some may be ruined by interference. The detector will collect over a range of scattering angle and this resolution may not be constant. Each step in the experiment can be studied. We can reject bad measurements and we are left with a set of data. From a conventional viewpoint this data may have 'gaps', but this is of no importance because, in our approach, we never go from data space to map space. We never process data!

The inverse operator that might take us from data to map space may not exist at all, usually operates on a space larger than the data space, is frequently unstable and may be very complicated. The philosophy put forward here is to consider all the solutions in map space that could give rise to the data. Since we can solve the forward problem, we can always check whether a solution is consistent, taking into account the errors in the data. Now we choose one of these solutions according to some criterion. If all the consistent maps are very similar, then there is no need to choose. There may be many different ones and then we need to invent a rule for choosing. I have described the maximum entropy choice and recommend it as a very good one.

References

Gull, S. F. (1989). Developments in maximum entropy data analysis. In *Maximum entropy and Bayesian methods, Cambridge, England, 1988* (ed. J. Skilling), pp. 53–71. Kluwer, Dordrecht.

Jaynes, E. T. (1986). Monkeys, kangaroos, and N. In *Maximum entropy and Bayesian methods in applied statistics: proceedings of the fourth maximum entropy workshop, University of Calgary, 1984* (ed. J. H. Justice), pp. 26–58. Cambridge University Press.

Jeffreys, H. (1961). *Theory of probability* (3rd edn). Clarendon Press, Oxford.

Skilling, J. (1989). Classic maximum entropy. In *Maximum entropy and Bayesian methods, Cambridge, England, 1988* (ed. J. Skilling), pp. 45–52. Kluwer, Dordrecht.

Skilling, J. and Bryan, R. K. (1984). Maximum entropy image reconstruction: general algorithm. *Monthly Notices of the Royal Astronomical Society*, **211**, 111–24.

2

Fundamentals of MaxEnt in data analysis

J. Skilling

Abstract

The maximum entropy method (MaxEnt) demonstrably produces good reconstructions of images and spectra from blurred or otherwise imperfect data. However, these reconstructions have traditionally been single selections, lacking any estimates of their accuracy. Recently, the application of MaxEnt has been placed on a clear axiomatic foundation. The entropy of an image is now seen to represent its prior probability. As a consequence of this deeper viewpoint, it is now possible to place error bars on MaxEnt reconstructions, so that the reliability of the features produced can be quantitatively assessed. It is also possible to assess different variants of MaxEnt.

2.1 Introduction

In data analysis, maximum entropy techniques are traditionally used to reconstruct positive distributions such as images and spectra from imperfect data. Very many distributions of practical interest, from nuclear physics to astronomy, are positive, and this gives the technique correspondingly wide applicability. By themselves, the subjective qualities of visual clarity and relative freedom from artefacts which characterize maximum entropy reconstructions would ensure the technique a permanent place in the repertoire of algorithms.

However, maximum entropy is not just *a* good technique for such problems: it is *the* technique. Latterly, we have found that maximum entropy rests on the firm foundation of probability calculus, which itself has a special status as the only internally consistent language of inference. Within this language, we find that positive distributions ought to be assigned probabilities which are based on the entropy of that distribution. The maximum entropy reconstruction corresponds to selecting the most probable individual one, because the entropy is thereby maximized. Around this, though, lies a probability distribution which allows us to place error bars on the reconstruction. For the first time, these enable us to assess quantitatively and objectively the reliability of the features which we see. Beyond this, we can assess different variants of maximum entropy, and give quantitative comparisons.

In this paper, we shall start at the beginning. Given imperfect data, what can we infer about the underlying image, spectrum, crystal structure, or whatever? This is a non-trivial problem of inference, because the image will be described by hundreds, thousands, or perhaps millions of numbers, and not just one or two. Difficult problems, though, have been solved before in other areas, and appropriate techniques have been developed. In engineering, it is known that difficult tasks require highly reliable components. For example, the few million components of a spacecraft have to be more reliable than the few hundred of a bicycle, the consequences of failure being all too obvious. Likewise in computing, we do not relax the rules of arithmetic when a calculation becomes long: instead we go to considerable trouble to ensure that the computation is carried out precisely, without intermediate error. So it should be in inference. Fuzzy thinking may suffice for simple problems, where the answers are reasonably obvious anyway, but we expect accurate logic to be needed for larger problems.

2.2 Probability calculus

Scientists simplify, having learned that complicated problems can be broken down into simpler ones. We will follow this methodology, and start at the beginning with general reasoning about simple situations. In all inference problems, we have to consider different propositions. In image reconstruction, these might relate to particular images: thus

> Proposition A: `the image is of an aardvark`,
> Proposition B: `the image is of a beagle`,
> Proposition C: `the image is of a cat`, etc..

At the very least, we will want to be able to rank our preferences for these different propositions, for example we might

> Prefer B to A AND Prefer A to C.

Presumably we want these to imply

> Prefer B to C,

otherwise we will soon start to argue in circles. In other words, we need a transitive ranking of preferences. Consequently, we can assign a real number to each preference, slotting in a new number for each new preference depending on where it falls in the existing transitive sequence. Although fairly arbitrary, these numerical codes

$$\pi(\text{Proposition}) \in \mathbb{R}$$

are at least correctly ordered. Let us apply this minimal structure to the simplest problems.

1. Start with a '1-bit' proposition $X \in \{\text{True, False}\}$. For example, $X =$ `Confucius was born in 551 BC`. X may or may not be true: we don't really know, and we can only express a preference, based, as well as may be, on such information as we have to hand. Presumably, though, our preference for X being False will be determined by our preference for X being True.

$$\text{Preference for } {\sim}X \Leftarrow \text{Preference for } X.$$

In terms of our numerical codes, this means that there is some mapping (i.e., a function) f, which takes our code for a proposition into the corresponding code for its negation.

$$\exists f : \pi({\sim}X) = f\big(\pi(X)\big).$$

Already, we can start doing some mathematics, because the identity ${\sim}{\sim}X = X$ immediately tells us that

$$f\big(f(x)\big) = x,$$

so that f is not wholly arbitrary. However, this does not take us very far.

2. Moving on, we take a 2-bit proposition $(U, V) \in \{\text{TT, TF, FT, FF}\}$. We can reach our joint preference for the two bits (U, V) being, say, TT (both true) in two individual 1-bit steps.

1. Express preference for U (say),
2. Only if U is true, obtain and use preference for V, given U.

In other words,

$$\text{Preference for } (U, V) \Leftarrow \text{Preference for } U, \text{ Preference for } (V|U).$$

In terms of the codes, there is some function g which generates this joint preference:

$$\pi(U,V) = g\big(\pi(U), \pi(V|U)\big).$$

Again, we can do a little mathematics, using the Boolean identity $(U,V) = (V,U)$ to reach

$$g\big(\pi(U), \pi(V|U)\big) = g\big(\pi(V), \pi(U|V)\big),$$

but the structure remains too impoverished to take us very far.

3. Hence we move on to a 3-bit proposition (R,S,T). This can be factored into simpler conditional propositions in half a dozen different ways, such as:

1. Express preference for R (say),
2. Only if R is true, obtain preference for S (say), given R,
3. Only if R and S are both true, obtain preference for T, given R and S.

All of these half a dozen ways must, of course, be equivalent, because the order of R, S and T is irrelevant to their joint truth. The truly remarkable consequence of this elementary observation (Cox 1946), is that there exists some function F of our original rather arbitrary numerical codes π, taking them into other codes

$$\mathrm{pr}(X) = F\big(\pi(X)\big)$$

which are distinctly less arbitrary. In fact the new codes obey

$$\boxed{\begin{array}{l} \mathrm{pr}(X) + \mathrm{pr}({\sim}X) = 1, \\ \mathrm{pr}(X,Y) = \mathrm{pr}(X)\,\mathrm{pr}(Y|X), \\ \mathrm{pr}(\mathrm{False}) = 0 \qquad \text{and} \qquad \mathrm{pr}(\mathrm{True}) = 1. \end{array}}$$

At this point or before, a purist might remember that any preference whatever must be conditional upon some sort of earlier expectation or belief, traditionally given the symbol I, so that we should more properly write

$$\mathrm{pr}(X|I) + \mathrm{pr}({\sim}X|I) = 1,$$
$$\mathrm{pr}(X,Y|I) = \mathrm{pr}(X|I)\,\mathrm{pr}(Y|X,I).$$

In the interests of notational clarity though, we will often omit this I, which may in any case be a compound proposition common to several steps in an argument.

Anyway, we recognize the standard rules of probability calculus, being used precisely for their original purpose of quantifying our preferences, and we are entirely correct to identify our new codes pr as probability values. They happen to obey the old-fashioned frequentist definition of probability when that is germane. Indeed, given an infinite sequence of trials, how else could one expect to define one's belief about an arbitrary individual success but on the basis of the overall success ratio? However, our interpretation of probability is far more general. To take the above example, Confucius presumably had only one birthdate, and not an infinite ensemble of them. We can, nevertheless, discuss it probabilistically with full propriety. To follow Jaynes (1989) in using the terminology of philosophy, *probabilities are epistemological*, representing our beliefs, rather than ontological, representing objective external reality.

The point being stressed here is not that we can describe our preferences by probability values. The point is that we *must* do so (or adopt an equivalent description, such as percentages in which all the codes are artificially multiplied by 100). The *only* language of inference which deals consistently with simple problems is ordinary probability calculus, just as originally required by Laplace (1814). Technically we have not proved that probability calculus is itself internally consistent, any more than that we have proved arithmetic to be consistent. All we know is that no different calculus can be consistent. If, though, any inconsistency is ever found, I for one intend to retire to my garden (if indeed a refuge from so fundamental a calamity could be found even there).

Any complicated proposition may be constructed from simpler ones, just as any complicated quantity in computer memory can be broken down into its constituent bits. Hence we must also use probability calculus when dealing with complicated problems. Indeed, the analogies with engineering and with computing suggest that strict submission to this discipline will become even more important as our problems become harder. Moreover, the analogies also suggest that the benefit of adhering to the discipline will be qualitative improvements in reliability, precision, and power, with benefits far outweighing the costs.

Those workers who do argue in strict probabilistic terms are called 'Bayesians', although it seems perverse to have a special adjective (especially as Bayes' writings were less clear and extensive than those of Laplace). Outsiders are thereby given the impression that the strict probabilistic approach is just one of several, competing with other general schools such as fuzzy logic (Klir 1987), or with specific schools such as generalized cross-validation (Golub *et al.* 1979) and many more. As a matter of history, that has too often been the case, and it continues to the present day, but as a matter of philosophy it should not be. Consistent reasoning demands probability calculus. Logic has spoken.

That said, I hold that there is a valid defence of using non-Bayesian methods, namely incompetence. One may not see how to set up an acceptable probabilistic analysis, or one may not see how to evaluate the relevant probabilistic formulae, and in such case one has no choice but to botch one's approach, even though the results will be damaged. If results are needed, it must surely be acceptable to do the best one can. I hope so, for I have myself been in this situation embarrassingly often in the past. Conversely, being Bayesian does not make one right. A Bayesian can make wrong or dubious assumptions, just like anybody else.

2.3 Image reconstruction

Returning to our initial problem of image reconstruction, we can start writing down some mathematics. Let D be our data, and let f be our image, drawn from some, usually large, space of possibilities. As a matter of policy, we recommend starting a probabilistic analysis with the joint distribution of everything relevant. So far, we have

$$\mathrm{pr}(f, D) = \mathrm{pr}(f)\,\mathrm{pr}(D|f) = \mathrm{pr}(D)\,\mathrm{pr}(f|D).$$

In this expression, we identify

$$\mathrm{pr}(f) \text{ as a `prior' probability of } f.$$

This is the probability distribution which one would assign (presumably on some theorist's advice) before acquiring one's data. We shall return to this later. We also have

$$\mathrm{pr}(D|f) = \text{`likelihood'}$$

which is the conditional probability of acquiring the particular data D, given f. The form of this ought to be provided by the manufacturer of the observing equipment, who should have calibrated it over sufficiently many input signals to have a reasonable idea of what the output D would be for any given input. Parenthetically, this is the exceptional case where a frequentist estimation of probability would be acceptable. Of course, the prediction of output values will usually be imprecise. We call this imprecision 'noise', and we have to allow for it in the form of $\mathrm{pr}(D|f)$.

A common idealization here is to a linear experiment with Gaussian noise, often written in shorthand as

$$D = Rf + \sigma n,$$

where R is the response matrix of the equipment, σ is the noise standard deviation, and n is drawn from the unit normal distribution. Our knowledge of n is described by the probability distribution

$$\mathrm{pr}(n) = (2\pi)^{-\frac{1}{2}} \exp(-n^2/2).$$

With N data D, this immediately gives us

$$\mathrm{pr}(D|f) = \left(\prod (2\pi\sigma^2)^{-\frac{1}{2}}\right) \exp\left(-\tfrac{1}{2}\chi^2\right)$$

where

$$\chi^2 = \sum \sigma^{-2}(D - Rf)^2$$

is the usual chi-squared misfit statistic.

Continuing our identification of terms,

$$\mathrm{pr}(D) = \text{‘evidence’},$$

which we choose to measure in decibels:

$$\text{evidence} \sim 10 \log_{10} \mathrm{pr}(D) \quad \text{decibels}.$$

Along with Jaynes (private communication), we choose decibels because most of us have a clearer intuitive understanding of powers of 10 than of powers of e, and decibels are already widely used in engineering disciplines.

The evidence, like other terms, is implicitly conditioned on the theorist who is advising the prior on f. Later, we will see how to use the evidence $\mathrm{pr}(D|\text{theorist})$ to discriminate between different theorists, but for the moment we will treat it as a mere scaling constant, independent of f. Its value is

$$\mathrm{pr}(D) = \int \mathrm{d}f \; \mathrm{pr}(f, D) = \int \mathrm{d}f \; \mathrm{pr}(f)\,\mathrm{pr}(D|f),$$

involving an integral or sum over all possible images f. This process of integrating out from the joint probability distribution such variables as one is not interested in at the time is called ‘marginalization’.

Finally,

$$\mathrm{pr}(f|D) = \text{probability of } f \text{ after observing } D.$$

This is what we initially sought. It quantifies our inferences about the image f. In a very real sense, $\mathrm{pr}(f|D)$ is our result.

If we had to produce just one single image as the ‘best’ reconstruction, we would presumably give the most probable one, which maximizes $\mathrm{pr}(f|D)$. This single image would, of course, be incomplete without some statement of reliability, derived from the spread of reasonably probable f. We are accustomed to placing error bars on measurements of single variables, and we should also give them on images. Obtaining proper quantitative error bars is one of the qualitative benefits of a fully probabilistic treatment. Another benefit, less expected but scarcely less important, is the evidence $\mathrm{pr}(D)$.

Disentangling our inference about f, we reach

$$\mathrm{pr}(f|D) = \mathrm{pr}(f)\frac{\mathrm{pr}(D|f)}{\mathrm{pr}(D)}.$$

This equation is commonly known as Bayes' theorem. And the Awful Truth is laid bare. The measurements which gave us $\mathrm{pr}(D|f)$ do not fully define our result. We also need to assign the prior $\mathrm{pr}(f)$. Probability calculus is a language which shows us how to modify our preferences in the light of experience. But it does not tell us what our initial preferences should be. The language does not tell us what to say.

A little reflection shows that this flexibility is in fact entirely appropriate, and we should not wish it away, but it leaves us in the present instance with a difficulty over dimensionality. If we were just measuring a single scalar variable, such as the length of a rod, we would usually be able to make fairly precise measurements, sharp enough effectively to override our presumably weaker prior ideas of how long the rod might be. Indeed, as the noiseless limit of the likelihood $\mathrm{pr}(D|f)$ (a delta function) is approached, the prior becomes irrelevant. So does probability calculus become irrelevant, because the datum forces the 'correct' result, almost regardless of theory.

Images, though, are not just scalar quantities. They are usually digitized on thousands or even millions of cells, each of which has an intensity to be estimated. We can't get away with sloppy reasoning (though most of us have been known to try, inadvertently or otherwise), and the prior $\mathrm{pr}(f)$ plays a crucial role.

2.4 The probability of an image

At first sight, it seems a daunting task to assign a prior probability over the very large space of possible images. Much detail comes to mind about the sorts of pictures we have seen in the past, and it is tempting to try to analyse a new image in terms of structures that we have already learned about from others. By and large, this is the approach adopted by the artificial intelligence community. Within a suitably restricted context, such as the positioning of previously defined objects, the approach has had its successes. At the time of writing, though, these successes have been less than overwhelming in more general contexts, and we prefer to start again at the beginning, with simple problems. After all, simple arguments have already defined our language of inference, which might initially have seemed a no less daunting problem.

Various arguments of widely differing style, approach and rigour have been used in the past (Gull and Daniell 1979; Shore and Johnson 1980; Tikochinsky *et al.* 1984; Levine 1986; Skilling 1988, 1989; Rodriguez 1989). There is always scope for simplifying the assumptions, and to that end we here develop a yet different approach.

At the outset, we know some general qualitative features of the image or spectrum that we seek. It is positive, $f \geq 0$ on each cell, and it is additive, meaning that

$$\sum_i f_i \qquad \text{or equivalently} \qquad \int f(x)\,\mathrm{d}x$$

represents the total quantity residing in the specified domain. For example, light intensity is positive and additive, its sum representing a physical energy flux. By contrast, the amplitude of (incoherent) light is not additive.

As a corollary of additivity, f is a distribution, which should appear only in the form of a linear integral

$$\int f(x)\ldots\mathrm{d}x.$$

We shall call integrals of this form, or functions of them, 'observables', to distinguish them from other non-observable forms such as $\int f^2\mathrm{d}x$, $\int f^{\frac{1}{2}}\mathrm{d}x$, $\int f \log f\,\mathrm{d}x$ and so on. Data D are observable, non-linear data merely being non-linear functions of integrals which are themselves linear in f. Likewise, any inference we draw from observations must also be observable, otherwise we would find ourselves measuring a breakdown of the distributive property. Remarkably, this alone suffices to determine our formulae.

We write our prior probability density as

$$\mathrm{pr}(f) = \mu(f)\exp\bigl(-H(f)\bigr),$$

explicitly separating the f-space measure μ from the pointwise probability, written for convenience as $\exp(-H)$. We shall require the prior to be normalized, so that

$$\int \mathrm{pr}(f)\,\mathrm{d}f = 1.$$

Suppose that we have N linear data of unit standard deviation of noise:

$$D = Rf \pm 1,$$

or, more precisely, using vector notation,

$$\mathrm{pr}(D|f) = (2\pi)^{-N/2} \exp\Big(-(Rf - D)^{\mathrm{T}}(Rf - D)/2\Big).$$

After all, we certainly want our formulae to apply in this particular case. Our joint probability distribution becomes

$$\mathrm{pr}(f, D) = (2\pi)^{-N/2} \mu(f) \exp\Big(-H(f) - (Rf - D)^{\mathrm{T}}(Rf - D)/2\Big).$$

Let $\hat{f}$ be the image which maximizes the exponent, being defined by

$$\nabla H = -R^{\mathrm{T}}(Rf - D),$$

where $\nabla H = \partial H/\partial f$. Given the data, this is the most probable individual image. Expanding about $\hat{f}$, the evidence integrates by steepest descents to

$$\begin{aligned} \mathrm{pr}(D) &= \int \mathrm{d}f \ \mathrm{pr}(f, D) \\ &= (2\pi)^{-N/2} \mu(\hat{f}) \exp\Big(-H(\hat{f}) - (R\hat{f} - D)^{\mathrm{T}}(R\hat{f} - D)/2\Big) \\ &\quad \times \Big(\det\big((\nabla\nabla H + R^{\mathrm{T}}R)/2\pi\big)\Big)^{-\frac{1}{2}}. \end{aligned}$$

Evidence, being an inference, is observable. Certainly the mock data

$$F = R\hat{f}$$

are observable. Hence the remaining factors

$$\mu(\hat{f}) \exp(-H(\hat{f})) \big(\det(\nabla\nabla H + R^{\mathrm{T}}R)\big)^{-\frac{1}{2}}$$

form an observable. Moreover, this is observable for (effectively) all R at any fixed $\hat{f}$, because (unless R happens to be inconveniently null) we can steer to the desired $\hat{f}$ by manipulating D appropriately. Purely as a formal mathematical device, perturb R, then set $R = 0$. The expression

$$\mathrm{trace}\big((\nabla\nabla H)^{-1} \delta(R^{\mathrm{T}}R)\big)$$

is observable for all perturbations δR. Let Ω diagonalize $\nabla\nabla H$, and set

$$\hat{f} = \Omega h.$$

We call Ω the 'hidden matrix' and h the 'hidden image'. The inverse curvature matrix

$$C = \left(\frac{\partial^2 H}{\partial h \partial h}\right)^{-1}$$

is now diagonal. Because the trace above is observable for all δR,

$$\sum_i C_i \rho_i = \int C(x)\rho(x)\,\mathrm{d}x$$

must be observable for arbitrary ρ. Hence the inverse curvature C must be a distribution in its own right. Being observable, it can only be linear in $\hat{f}$, and hence linear in h also. However, because the inverse curvature is diagonal, each C_i can depend only on h_i, and not on any other h_j. Thus C_i is linear in h_i and we reach the central result of this derivation:

$$\left(\frac{\partial^2 H}{\partial h_i^2}\right)^{-1} = \frac{h_i - z_i}{w_i},$$

this being the most general linear form. We now integrate this to determine H itself. Continuing to choose notation for the arbitrary constants which aids their identification, we reach

$$\frac{\partial H}{\partial h_i} = w_i \log\left|\frac{h_i - z_i}{m_i - z_i}\right|$$

and

$$H = K - \sum_i w_i\left(h_i - m_i - (h_i - z_i)\log\left|\frac{h_i - z_i}{m_i - z_i}\right|\right).$$

In these expressions, we identify symbols as follows:

z_i = zero level = minimum value for h_i (if h is bounded below, or maximum if bounded above). With a positivity constraint in our images, it is natural to set $z = 0$, and henceforward we shall do so. It is easy to replace z in the equations if needed. The limit $z = -\infty$ is also of interest, corresponding to the absence of any positivity constraint.

m_i = 'model' or measure on coordinate space. In the absence of any other constraint, H is minimized at $h = m$.

w_i = weight on cell i. Occasionally it is useful to keep individual cell-by-cell weights, but they can always be absorbed into the columns of Ω, so we shall set $w_i = \alpha$ = 'regularization' constant. Again, w is easy to replace if needed. Actually, we could absorb α too into Ω, but that proves less convenient.

K = normalizing constant, which we set to 0 because we still have freedom to rescale the measure μ to compensate.

In summary, we have

$$H(\hat{f}) = -\alpha S(h),$$

where

$$\hat{f} = \Omega h \qquad \text{with} \qquad h = \text{hidden image}$$

and

$$S(h) = \sum_i \big(h_i - m_i - h_i \log(h_i/m_i)\big),$$

which we recognize as the entropy of the (hidden) image, in the form given by Skilling (1988). Indeed, without proceeding any further, we have a justification for 'maximum entropy' data analysis. The most probable individual image is that which maximizes the entropy, subject to appropriate data constraints.

Notwithstanding its apparently nonlinear form, the entropy S (or equivalently its scaled form H) is observable, because

$$\alpha \log(h/m) = \partial H/\partial h = \Omega^{\mathrm{T}} \partial H/\partial f = -\Omega^{\mathrm{T}} R^{\mathrm{T}} (Rf - D).$$

Substituting this logarithm in S yields the alternative expression

$$S(\hat{f}) = \sum (h - m) + F^{\mathrm{T}}(F - D)/\alpha,$$

which is clearly observable.

However, our analysis is not yet complete. We still need

$$\mu(\hat{f})\big(\det(\nabla\nabla H + R^{\mathrm{T}}R)\big)^{-\frac{1}{2}}$$

to be observable for all R. In particular, when $R = 0$,

$$\log \mu(\hat{f}) - \tfrac{1}{2} \log \det(\nabla\nabla H) = \log \mu(\hat{f}) - \tfrac{1}{2} \sum_i \log(\alpha/h_i) + \tfrac{1}{2} \log \det(\Omega\Omega^{\mathrm{T}})$$

must be observable. Hence we arrive at the measure

$$\mu(\hat{f}) = Q\big(\det(\Omega\Omega^{\mathrm{T}})\big)^{-\frac{1}{2}} \prod_i (\alpha/h_i)^{\frac{1}{2}},$$

where Q is observable. In fact, Q is the coefficient $(2\pi)^{-M/2}$ needed to ensure that the prior is properly normalized (over M cells). Our main task is complete: we have our prior on images. It has turned out to pertain to a hidden image h, and is really quite simple:

$$\boxed{\mathrm{pr}(h|\alpha, \Omega, m) = \Big(\prod_i (\alpha/2\pi h_i)^{\frac{1}{2}}\Big) \exp\big(\alpha S(h)\big).}$$

The data relate to this through

$$D = \tilde{R}h \qquad \text{with} \qquad \tilde{R} = R\Omega$$

and give us the evidence in the form (after a little symbol-shuffling)

$$\begin{aligned}\mathrm{pr}(D|\alpha,\Omega,m) &= (2\pi)^{-N/2}\Big(\det\big(I + \tilde{R}\,\mathrm{diag}(h)\tilde{R}^{\mathrm{T}}/\alpha\big)\Big)^{-\frac{1}{2}} \\ &\quad \times \exp\big(\alpha S(h) - (\tilde{R}h - D)^{\mathrm{T}}(\tilde{R}h - D)/2\big),\end{aligned}$$

where h maximizes the exponent. We also obtain the probability distribution of h surrounding the maximizing value, which in the Gaussian approximation is

$$\mathrm{pr}(h|D,\alpha,\Omega,m) \propto \exp\Big(-\delta h^{\mathrm{T}}\big(\alpha\,\mathrm{diag}(h^{-1}) + \tilde{R}^{\mathrm{T}}\tilde{R}\big)\delta h/2\Big).$$

From this we can derive probabilistic estimates of our 'real' image f, and of any of its features. Correctly, such features are all observable.

2.5 The quantitative assessment of theories

In order to use these powerful results, we need to specify three quantities, the hidden matrix Ω, the spatial model m, and the regularization constant α. It seems that one is forced to argue about positive additive images in these terms, though different theorists might legitimately differ in their views on Ω, m and α. Of these, the hidden matrix Ω is perhaps the most crucial, because it defines the very structure of the problem.

Most simply, Ω could be taken to be the identity matrix. We call this the 'classical' viewpoint, and personalize it by attributing it to a theorist called 'Mr Classic'. Mr Classic believes that all the cells of his image are independent, so that there are no correlation terms in his prior on the real image f. Hence his Ω is diagonal, all its diagonal elements being made equal by the usual assumption of spatial invariance. Any scaling of Ω is incorporated in α, so Mr Classic ends up by identifying h with f itself. Historically, the great majority of work in maximum entropy data analysis has followed this identification, in which the entropy is placed directly on the actual image f.

However, alternatives are certainly possible. For example, Ω could be a convolution matrix, with

$$f(x) = b(x) * h(x) \qquad \text{with} \qquad b = \text{preblur function.}$$

This alternative, personalized by ascribing it to 'Mr Preblur', has the hidden image blurred by convolving it with some 'preblur' function, possibly a

Gaussian of some specified width b. Charter (unpublished thesis, 1989) has used this technique in pharmacokinetics. Of course, it would be incumbent upon Mr Preblur to specify the degree of blurring in order to define Ω fully, which he must do by ascribing to it his personal probability distribution $\mathrm{pr}(b|\,\text{Mr Preblur})$.

Next, we have to specify the model $m(x)$. Again, different theorists may have different prejudices about m, though many of them would agree that in the absence of any prior information favouring some particular coordinates x over others, m should be just a scalar, independent of x. Again, though, it must be specified in advance with its own individual prior $\mathrm{pr}(m|\,\text{theorist})$. Often enough, one just fixes m at some *a priori* plausible value. For example, astronomers might (should?) fix m to be the average sky brightness, to which value their images should default in the absence of any relevant data.

The remaining quantity α is also a scalar. This dimensional parameter is unlikely to be known in advance, and it seems sensible to ask that it be assigned some broadish prior, flattish in the logarithm. The precise details should not matter too much.

At this stage, our theorists are giving us families of priors for our image, each member being defined by a particular choice of just a few scalars (specifically α, m and possibly b). They are also giving us their prior distributions for those scalars. Thus we are being given formulae for the (hidden) image, as inferred from the data,

$$\mathrm{pr}(h|D,\,\text{scalars, theorist})$$

and for the evidence

$$\mathrm{pr}(D|\,\text{scalars, theorist}),$$

as well as the prior expectation

$$\mathrm{pr}(\text{scalars})$$

of what those scalars should be. Now we are not usually very interested in the scalars, so we should integrate them out of the joint distribution, obtaining

$$\mathrm{pr}(h|D,\,\text{theorist}) = \int d(\text{scalars})\,\mathrm{pr}(\text{scalars}\,|\,\text{theorist})\,\mathrm{pr}(h|D,\,\text{scalars, theorist})$$

and

$$\mathrm{pr}(D|\,\text{theorist}) = \int d(\text{scalars})\,\mathrm{pr}(\text{scalars}\,|\,\text{theorist})\,\mathrm{pr}(D|\,\text{scalars, theorist}).$$

Actually, one can in practice short-circuit these integrals over scalars. In large-scale problems, with hundreds or thousands of data available, the posterior distribution of any scalar quantity is likely to be sharply peaked around its 'best' most probable value. Hence the integral can be evaluated to within a relatively unimportant factor of order unity by simply taking the integrand at the maximum:

$$\int \cdots \mathrm{d}(\text{scalar}) = [\cdots] \qquad \text{at maximizing scalar.}$$

At last, all the theory is set up. The arguments concerning observables have been occasionally intricate, but that is because we have wished to start from first principles, and expose the generality and power of this quantified maximum entropy technique from minimal assumptions. We are at the point where we can start to analyse some data.

As a philosophical aside, it is interesting to return here to the question of competence. In order to obtain the desired results, one has to face the problem of writing computer programs to perform the specified integrals, evaluate the determinants, and so on, all in very large spaces indeed. Myself, I believed for too long that the required matrix operations in f-space were far too large to handle, so that I was unable to use, and had little interest in helping to develop, this theory. The language of probability calculus offers no guidance to those who can't do the sums. Anyway, I was wrong. We could and did write the programs.

2.6 Deconvolution of Poisson data

Deconvolution of a blurred and reasonably noisy data set forms a good, standard, test bed for any image reconstruction technique. The results are directly visual, so that the qualitative strengths and weaknesses of a technique can easily be illustrated. Often enough, a digitized optical photograph is used as an example, and usually the data are linear with Gaussian noise. Sometimes indeed, as with quadratic Wiener filters, a reconstruction technique demands these assumptions. In order to be a little different, and to demonstrate the generality of the probabilistic approach, we shall discuss a blurred data set subject to Poisson statistics. Specifically, we take a set of gamma-ray counts from a radioactive source as obtained from a medical camera. The data (Fig. 2.1, courtesy of Stephen Brown at Southend Hospital) were actually of a thyroid phantom. The instrumental point-spread function, straightforwardly obtained as the response to a point source, was also supplied (Fig. 2.2).

These data intensities ranged downwards from 100 or so in the brightest of the 64×64 data cells to a dim background containing only a few stray background counts. Clearly, it would have been illegitimate to approximate

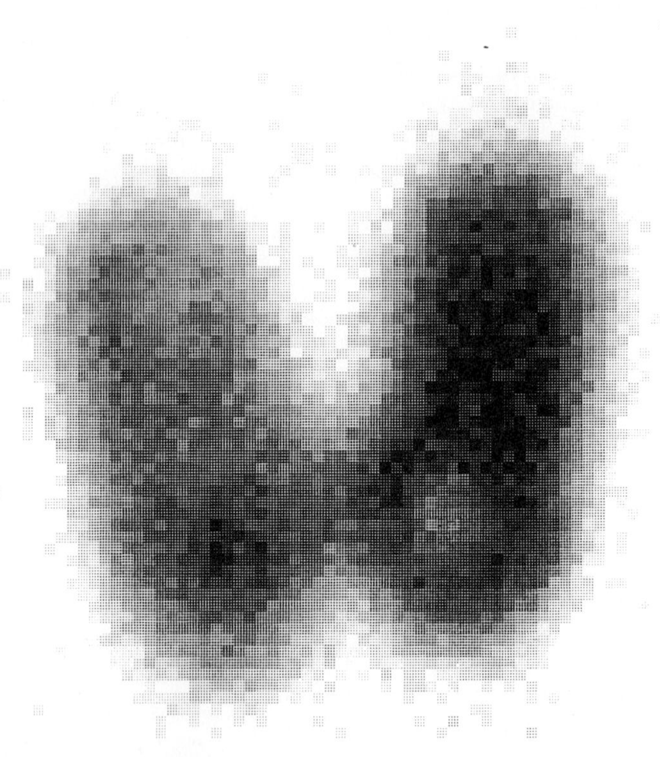

Fig. 2.1. Data. Gamma-ray tomogram of a thyroid phantom on 64×64 cells (S. Brown, Southend Hospital).

Fig. 2.2. Point-spread function, from a gamma-ray point source.

such data with a Gaussian likelihood function, so the proper Poisson form for integer data D with expectations F was used, namely

$$\text{likelihood} = \text{pr}(D|F) = \prod_k F_k^{D_k} \mathrm{e}^{-F_k} / D_k!.$$

The entropic prior is, of course, unaffected by this change, and the 'best' image remains defined as that which maximizes the exponent

$$\alpha S(f) + \log\big(\text{likelihood}(f)\big).$$

We still need to approximate the posterior probability distribution $\text{pr}(f|D)$ around this maximum by a local Gaussian form, in order to define the error bars comprehensibly and to do the computations, but this should be less damaging than approximating the likelihood globally in advance.

On integrating over f, Mr Classic's prior gave evidence of

$$\text{pr}(D|\,\text{Classic}) = \exp(-9736) \qquad \text{or} \qquad \text{evidence} = -42\,283\,\text{db}.$$

One might think that this number is rather small, but something of the sort should be expected. After all, there are a thousand or more cells with counts of 100 or so, so that any particular set of integer counts will be rather improbable. This is reflected in the numerical value of the evidence.

Mr Classic's 'best' image (Fig. 2.3), though, is absolutely dreadful. It is appallingly 'spotty', with its brightest cells ranging beyond 1000 in magnitude, ten times brighter than any individual data. The deconvolution has gone far beyond what any user should, and what any experienced user would, desire. In attempting to sharpen up the image by removing the blurring imposed on the data by the point-spread function, the program has produced a picture in which the noise has been amplified to an intolerable degree.

As it happens, Poisson data and maximum entropy together are particularly prone to this behaviour. Poisson data have larger absolute errors on the more intense data values, varying as the square root, so that if noise amplification is to occur, it will be seen first on the bright regions. The entropy formula, dominated as it is by its '$f \log f$' term, is relatively tolerant of bright spots of large f, as compared with, say, its old-fashioned quadratic rivals, so that it is relatively easy for such spots to develop. That is not necessarily a bad thing if the field of view contains just a few scattered spots, but this particular object contained reasonably extensive bright regions, allowing plenty of room for oscillations to develop. For all these reasons, this gamma-ray data set formed a particularly hard test. Nevertheless, the test is fair, because the data set is quite typical of its genre, and its genre is important. Many observers look at distributed objects by counting photons, and sometimes those photons are quite sparse.

Fig. 2.3. Reconstruction by Classic MaxEnt.

Actually, the results are not quite as bad as they look. The quantified error bars show that the excessively bright spots are not statistically significant. Typically, an individual bright cell quantifies as 1000 ± 1000, and it is only after several neighbours are averaged together that the intensities become significant. The probability formulae themselves show us that the wild oscillations are unreliable. Still, it is disconcerting to have to mentally defocus our reconstruction before our brains can understand it reasonably well, especially as such blurring will inevitably destroy some real information along with the amplified noise. Indeed, if quantified maximum entropy were entirely restricted to Mr Classic's assumptions, it would surely deserve to be rejected.

Fortunately, the theory allows us to decouple our actual image f from an underlying hidden image h, by using the hidden matrix Ω. If, as suggested above, we take Ω to be a convolution operation, then necessarily the actual 'best' image $\hat{f}$ which we generate will exhibit a corresponding degree of spatial correlation. Fairly abrupt local changes can still occur, but they

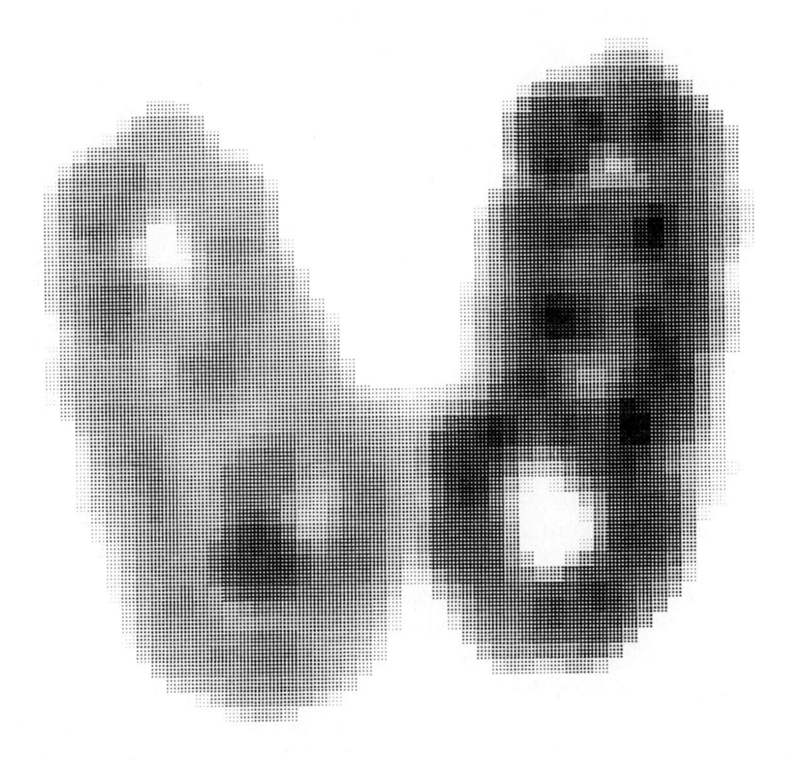

Fig. 2.4. Reconstruction by Preblur MaxEnt. The preblur was a circular disc of radius 3 cells. This reconstruction is more probable than the Classic reconstruction of Fig. 2.3 by a factor of 2×10^{77}.

have to be driven by very much more violent changes in the hidden image, and we may expect them to be discriminated against. Imposing spatial correlation through a suitably chosen hidden matrix is a gentle and sympathetic technique, and we may be grateful that the theory of quantified maximum entropy encourages us to use it.

Mr Preblur's 'best' reconstruction is shown in Fig. 2.4. (We choose not to display his hidden image, which looks truly dreadful and certainly ought to remain hidden.) Fig. 2.4 accords much more closely with what we might like to see. It is much easier to understand, and one immediately picks out two broad lobes, the right hand one brighter than the left, each with one significant hole, together which other irregularities which are more difficult to assess without experience. Mr Preblur's probabilistic error bars, though, are quite believable. Individual cells are not too reliable, as one might expect in a deconvolution, but integrals over 3×3 cell patches give sensible results, affording adequate discrimination between different areas.

Of course, one should not expect these error bars to be taken as absolute truth. They remain conditional on Mr Preblur's optimal hidden blur, of a circular disc of radius 3 cells, and one might wish to retain a little scepticism about this. Even if they were 'correctly' distributed, in the very nature of things about 1 in 20 of one's estimates would be discordant at the 95% significance level. Perhaps, though, one should not cavil. It is surely good that one can discriminate a pathology in the image from a pathology in the algorithm. Especially in medicine.

So far, we have argued on intuitive grounds that we like Mr Preblur's picture better than Mr Classic's, but what of the evidence? Mr Preblur's evidence evaluates to

$$\mathrm{pr}(D|\,\mathrm{Preblur}) = \exp(-9558) \qquad \text{or} \qquad \text{evidence} = -41\,510\,\mathrm{db},$$

which we may compare with Mr Classic's evidence of $\exp(-9736)$. The evidence numbers before us look like, and indeed are, likelihood factors for the theorist in question, after we have obtained the data. Quite simply, the data are measuring the theorist.

When a given theorist was estimating images, he used the likelihoods $\mathrm{pr}(D|f)$ to modify his priors $\mathrm{pr}(f)$ into posterior inferences $\mathrm{pr}(f|D)$. We can do exactly the same here. Acting as an unusually objective Examiner, we can use the evidences $\mathrm{pr}(D|\,\mathrm{theorist})$ to modify our prior $\mathrm{pr}(\mathrm{theorist})$ into posterior inferences $\mathrm{pr}(\mathrm{theorist}\,|D)$. There is nothing new in principle here: Harold Jeffreys (1939, Chapter 5) explained it half a century ago. According to the probability product law, we obtain

$$\frac{\mathrm{pr}(\mathrm{Preblur}\,|D)}{\mathrm{pr}(\mathrm{Classic}\,|D)} = \frac{\mathrm{pr}(\mathrm{Preblur})}{\mathrm{pr}(\mathrm{Classic})}\frac{\mathrm{pr}(D|\,\mathrm{Preblur})}{\mathrm{pr}(D|\,\mathrm{Classic})}.$$

In order to be initially fair to both, we might assign them equal prior probability. Our conclusion would then be that

$$\frac{\mathrm{pr}(\mathrm{Preblur}\,|D)}{\mathrm{pr}(\mathrm{Classic}\,|D)} = \exp(9736 - 9558) \text{ or } 773 \text{ decibels.}$$

That is a significant factor. We have gained in objective and quantified terms by allowing for some spatial correlation. It remains possible in principle to favour Mr Classic over Mr Preblur, but one's prior would have to be biased that way by more than 77 orders of magnitude. Most reasonable observers would accept that Mr Preblur was better, at least in this case.

Of course, the story need not end here. Some new competitor, with a different hidden matrix, might come on the scene, and outclass Mr Preblur as effectively as he outclassed Mr Classic. Nothing in the theory tells us to restrict our attention to convolution matrices. Indeed, looking to the future one may guess that the most successful hidden matrices will prove to be those which extract 'real' features of the image cleanly and without side-effects. Maybe this is where quantified maximum entropy will start to link up with artificial intelligence. Until such competitors arrive, though, we can't tell, because we can't explore the entire space of hidden matrices. We have to wait upon a proposal.

2.7 Conclusions

Four main points are made in this chapter.

1. We have to argue in the language of probability calculus when attempting to describe our preferences, and make inferences. That is a very wide-ranging matter, going far beyond image reconstruction. It imposes upon us the necessity of assigning a prior probability distribution before we can legitimately draw inferences from our data.
2. In image reconstruction, very general arguments lead us to construct an entropic prior, with a defined measure, on the space of possible images. There is, however, a twist in the argument. The entropy is not necessarily placed directly upon the image we see. Instead, it is placed upon a hidden image, from which we derive the actual 'real' image by an arbitrary linear transformation. This extra freedom can be used constructively, to considerable effect.
3. The fully quantified probabilistic analysis lets us assess different theories, as long as they are cast in probabilistic terms. Hence we have objective grounds for rejecting inferior approaches, and for assessing new ones.
4. We can obtain error bars on our reconstructed images, thus ascertaining whether or not any perceived feature is likely to be significant.

References

Charter, M. K. (1989). Maximum entropy pharmacokinetics. Ph. D. thesis. University of Cambridge.

Cox, R. T. (1946). Probability, frequency and reasonable expectation. *American Journal of Physics*, **14**, 1–13.

Golub, G. H., Heath, M., and Wahba, G. (1979). Generalized cross-validation as a method for choosing a good ridge parameter. *Technometrics*, **21**, 215–24.

Gull, S. F. and Daniell, G. J. (1979). The maximum entropy method. In *Image formation from coherence functions in astronomy* (ed. C. Van Schooneveld), pp. 219–25. Reidel, Dordrecht.

Jaynes, E. T. (1989). Clearing up mysteries—the original goal. In *Maximum entropy and Bayesian methods, Cambridge, England, 1988* (ed. J. Skilling), pp. 1–27. Kluwer, Dordrecht.

Jeffreys, H. (1939). *Theory of probability.* Clarendon Press, Oxford.

Klir, G. J. (1987). Where do we stand on measures of uncertainty, ambiguity, fuzziness and the like? *Fuzzy Sets and Systems*, **24**, 141–60.

Laplace, P. S. (1814). *Essai philosophique sur les probabilitiés.* Courcier Imprimeur, Paris.

Levine, R. D. (1986). Geometry in classical statistical thermodynamics. *Journal of Chemical Physics*, **84**, 910–6.

Rodriguez, C. (1989). The metrics induced by the Kullback number. In *Maximum entropy and Bayesian methods, Cambridge, England, 1988* (ed. J. Skilling), pp. 415–22. Kluwer, Dordrecht.

Shore, J. E. and Johnson, R. W. (1980). Axiomatic derivation of the principle of maximum entropy and the principle of minimum cross-entropy. *IEEE Transactions on Information Theory*, **IT-26**, 26–39 and **IT-29**, 942–3.

Skilling, J. (1988). The axioms of maximum entropy. In *Maximum entropy and Bayesian methods in science and engineering*, Vol. 1 (ed. G. J. Erickson and C. R. Smith), pp. 173–87. Kluwer, Dordrecht.

Skilling, J. (1989). Classic maximum entropy. In *Maximum entropy and Bayesian methods, Cambridge, England, 1988* (ed. J. Skilling), pp. 45–52. Kluwer, Dordrecht.

Tikochinsky, Y., Tishby, N. Z., and Levine, R. D. (1984). Consistent inference of probabilities for reproducible experiments. *Physical Review Letters*, **52**, 1357–60.

3

Maximum entropy and nuclear magnetic resonance

P. J. Hore

Abstract

The use of the maximum entropy method as an alternative to Fourier transformation in nuclear magnetic resonance (NMR) spectroscopy is described. The problems of sensitivity, resolution, truncation and dispersion lineshapes are addressed and examples of maximum entropy processing of two-dimensional NMR data are presented. Finally a new approach to the maximum entropy problem in NMR is outlined.

3.1 Introduction

Nuclear magnetic resonance (NMR) spectroscopy is an enormously powerful and versatile physical method for investigating the structure and dynamics of molecules. Over the last fifteen years or so, a vast array of ingenious and specialized NMR techniques has been devised to allow physicists, chemists, biochemists and clinicians to investigate systems as intricate as the human body, to determine the complete three-dimensional structure of molecules as complex as proteins, and to study processes on timescales from picoseconds to days. Despite all this sophistication and diversity, the numerical methods generally used to convert experimental data into interpretable spectra are rather basic and rely almost exclusively on a single, straightforward operation, the Fourier transform. In the following, I discuss a more powerful technique of data processing, the maximum entropy method, and indicate how and where it may be of use to practising NMR spectroscopists.

3.2 Nuclear magnetic resonance

NMR spectroscopy uses nuclear magnetism as a 'spy' to monitor and report on the structures, motions and behaviour of molecules. Most elements have

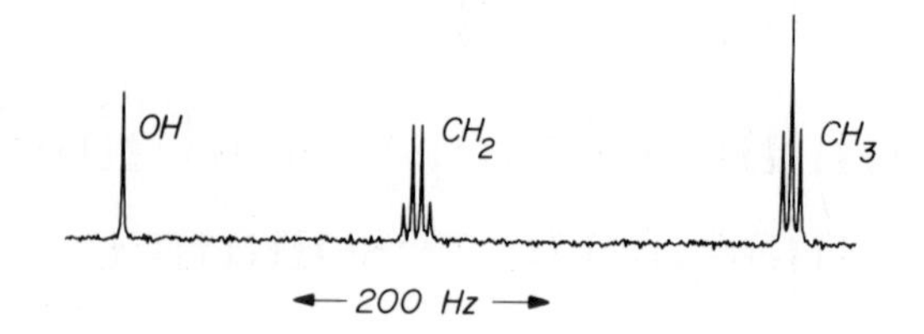

Fig. 3.1. ^{1}H NMR spectrum of liquid ethanol, CH_3CH_2OH.

at least one naturally occurring magnetic isotope, so that essentially every molecule one might wish to investigate has one or more spies already in place. Furthermore, these spies interact so weakly with their surroundings that they have a negligible effect on the properties of the molecules they probe.

A nucleus with a magnetic dipole moment has an associated intrinsic angular momentum known as spin. Both properties are quantized, such that a magnetic nucleus (a 'spin') in a magnetic field adopts one of a small number of orientations, with energies that depend on the local magnetic environment of the nucleus and contain information on the electronic structure of the molecule and the arrangement of nearby nuclei. NMR spectroscopy probes these energies using coherent electromagnetic radiation. Transitions from one spin orientation to another occur and are detected when the energy of the radiation matches the energy separations of the spin states, that is when the electromagnetic field is in resonance with the spins.

The basic features of NMR spectra of liquid samples are conveniently illustrated by the ^{1}H spectrum of liquid ethanol (Fig. 3.1). The spectrum, a plot of intensity against frequency, comprises three groups of peaks corresponding to the three chemically distinct types of proton in the molecule. The integrated intensities of the three groups, or multiplets, are in the ratio $1:2:3$ reflecting the number of hydroxyl (OH), methylene (CH_2) and methyl (CH_3) protons in the molecule. This discrimination between different nuclear environments is known as the chemical shift. The structure of each multiplet—singlet, quartet and triplet—arises from the interaction between neighbouring protons. The observed line-widths are determined partly by intrinsic spin relaxation processes and partly by instrumental factors.

Nowadays, NMR signals are recorded almost exclusively in the time domain. The sample is placed in a strong static magnetic field and is subjected to a pulse (or several pulses) of intense monochromatic radiofrequency radiation which excites all spins of a particular type, for example all protons. The spins then precess around the direction of the magnetic field (the z axis) each at its own resonance frequency, determined by its chemical shift and coupling to nearby nuclei. The x and y components of this oscillating magnetization are recorded over a period of several seconds (typically) as the spins relax back to equilibrium. Thus, the experimentally measured NMR signal (the free induction decay) is a superposition of damped sinusoidal oscillations of different frequencies, amplitudes and phases contaminated, inevitably, by noise arising from the electronic components of the NMR receiver. Although one could, in principle, interpret the free induction decay directly, the information it contains is more readily appreciated if presented in the form of a spectrum—hence the need for data processing.

This cursory description of NMR does not begin to do justice to the subject. For clear and authoritative expositions of modern NMR spectroscopy, the reader is referred to recent monographs by Harris (1983), Derome (1987), Ernst *et al.* (1987) and Freeman (1987).

3.3 Conventional data processing

The mathematical operation generally used to unravel the assorted oscillating components in a free induction decay is the Fourier transform. Consider, as a prototype NMR signal, an exponentially damped sinusoid with real and imaginary parts proportional to the x and y components of nuclear magnetization:

$$s(t) = \exp(\mathrm{i}\Omega t)\exp\bigl(-t/T_2\bigr). \tag{3.1}$$

Here Ω is the resonance frequency; T_2, the spin–spin relaxation time, characterizes the decay of x and y magnetization. (The analogous time constant for the return to equilibrium of the z component is known as the spin–lattice relaxation time, T_1.) Such a signal might arise from a single radiofrequency pulse, applied at $t = 0$. Let us assume that $s(t)$ is recorded from $t = 0$ out to some time much longer than T_2 (Fig. 3.2). Conventional processing (reviewed by Lindon and Ferrige (1980)) involves Fourier transforming $s(t)$ to obtain the spectrum $S(\omega)$:

$$S(\omega) = \int_{-\infty}^{\infty} s(t)\exp(-\mathrm{i}\omega t)\,\mathrm{d}t. \tag{3.2}$$

To evaluate this integral, one customarily assumes that $s(t)$ is zero for $t < 0$ (before the pulse), to obtain:

$$S(\omega) = \int_0^{\infty} \exp\bigl(\mathrm{i}(\Omega-\omega)t - t/T_2\bigr)\,\mathrm{d}t = A(\omega) + \mathrm{i}D(\omega), \tag{3.3}$$

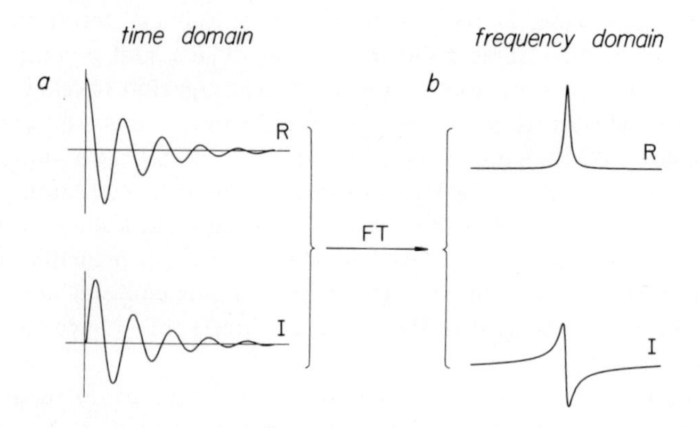

Fig. 3.2. Schematic time domain NMR signal (a) and its corresponding spectrum (b) obtained by conventional data processing. The real (R) and imaginary (I) parts of the free induction decay correspond to the x and y components of the precessing magnetization. The real and imaginary parts of the spectrum consist, in this simple example, of absorption and dispersion lines, respectively. FT denotes Fourier transformation.

where $A(\omega)$ and $D(\omega)$ are, respectively, absorption and dispersion mode lineshapes, centred at $\omega = \Omega$:

$$\begin{aligned} A(\omega) &= \frac{T_2}{1+(\Omega-\omega)^2T_2^2} \\ D(\omega) &= \frac{(\Omega-\omega)T_2^2}{1+(\Omega-\omega)^2T_2^2}. \end{aligned} \tag{3.4}$$

Conventionally one discards the imaginary part of the spectrum (the Hilbert transform of the real part) retaining the real part with its absorption mode lineshape. Note that a continuous Fourier transform has been used here for convenience: in reality the signal $s(t)$ is digitized at equal intervals and is processed by discrete Fourier transformation. Note also that the linearity of the Fourier transform ensures that equations (3.1)–(3.4) can be trivially extended to free induction decays comprising any number of superimposed oscillations. For more on the Fourier transform and its properties, see the books by Bracewell (1965) and Brigham (1974).

Two other procedures are routinely used in conjunction with Fourier transformation—digital filtering and zero-filling. Some form of filtering is almost always used to improve sensitivity or resolution and is most conveniently achieved by multiplying the free induction decay, point-by-point, by a suitable weighting function prior to Fourier transformation. Further discussion of this process is deferred to Sections 3.5.1 and 3.5.2. Zero-filling means supplementing the digitized free induction decay with a number of zeros so as to increase its length usually by a factor of two or more. The purpose of this is threefold: to improve digital resolution in the spectrum (that is, the number of points per unit frequency); to ensure that the real and imaginary parts of the spectrum are indeed related by Hilbert transformation (Bartholdi and Ernst 1973); and to permit the use of fast Fourier transform software which usually requires the number of data points to be a power of two.

To summarize, the Fourier transform is ubiquitous in NMR spectroscopy. It is a fast, convenient and (generally) trouble-free way of converting time domain data into frequency domain spectra. It is linear (so that the relative intensities of resonances of different widths, shapes and frequencies are not distorted) and is not model-dependent. Commercial NMR spectrometers come equipped with dedicated computers, array processors and fast Fourier transform software. Typically it takes a few seconds to transform a free induction decay of several thousand data points.

3.4 Maximum entropy

The maximum entropy method offers a very different approach to the processing of time domain NMR data. A clear account of the philosophy of the method and the fundamental arguments for using it are given in Chapter 1, and will not be repeated. Here, I shall briefly outline how maximum entropy may be applied to NMR data. (See Stephenson (1988) for a recent review.)

The fundamental disadvantage of Fourier transformation is that any defects in the data are simply transferred into the frequency domain along with the genuine signals. A more intelligent approach, upon which the maximum entropy method is based, is to proceed in the opposite direction—that is from the frequency domain to the time domain.

The principle (but not the practice) of the method may be summarized as follows. Suppose we were to guess a large number of trial spectra and subject them all to *inverse* Fourier transformation. We could immediately reject the vast majority of these trial time domain signals on the grounds that they bore no resemblance whatsoever to the experimental free induction decay. There would, however, be some that matched the data reasonably closely, and in the absence of extra knowledge about the spectrum, there would be no reason to prefer one of these trial spectra over the others. Faced with this range of spectra, all of which are compatible with the experimental data, the safest option is to pick the one with the minimum structure or equivalently the maximum entropy. Any other choice must be biased. Thus one seeks the spectrum with the maximum entropy from among those consistent with the data.

The most commonly used entropy is

$$S = -\sum_{n=1}^{N_\mathrm{f}} \left(\frac{x_n}{b}\right) \ln\left(\frac{x_n}{b}\right); \tag{3.5}$$

where x_n is the (real, positive) NMR intensity at the nth point in a trial spectrum which is digitized at N_f regular intervals and where b is a default parameter. The need for the trial spectrum to be compatible with the experimental signal may be stated in terms of a quantity χ^2 which should equal the number of data points $2N_\mathrm{t}$ ($= N_\mathrm{t}$ real $+ N_\mathrm{t}$ imaginary):

$$\chi^2 = \frac{1}{\sigma^2} \sum_{m=1}^{N_\mathrm{t}} \left|y_m - d_m\right|^2 = 2N_\mathrm{t}, \tag{3.6}$$

where d_m is the mth complex intensity in the experimental free induction decay, y_m is the mth complex intensity in the trial free induction decay and σ is the root-mean-square noise amplitude, assumed to be the same for all d_m. The idea is to match the y_m to the d_m to within a certain tolerance specified by the noise level. In this way one hopes to avoid introducing noise into the spectrum.

The trial free induction decay is normally obtained from the trial spectrum by a linear transformation:

$$y_m = \sum_{n=1}^{N_\mathrm{f}} R_{mn} x_n \qquad \text{or} \qquad \boldsymbol{y} = \mathbf{R}\boldsymbol{x}, \tag{3.7}$$

where $\boldsymbol{x}$ and $\boldsymbol{y}$ are column vectors and $\mathbf{R}$ is a matrix which is independent of $\boldsymbol{x}$. Often $\mathbf{R}$ is simply an inverse Fourier transform, but it may also include a time domain weighting function (for example, to produce resolution enhancement: see Section 3.5.2).

The task of maximizing S with respect to the x_n, where $n = 1, 2, \ldots, N_\mathrm{f}$, subject to the constraint $\chi^2 = 2N_\mathrm{t}$, is not as formidable as it sounds: in the N_f-dimensional space of the spectral intensities, the entropy is everywhere convex and its intersection with $\chi^2 = 2N_\mathrm{t}$ is a convex function with a single maximum. There are no local extrema (see Fig. 3.3). The constrained maximum may be located by an iterative search procedure (Skilling and Bryan 1984). Several search vectors are first chosen with directions related to the gradients of S and χ^2 with respect to the x_n. The search direction for each iteration is a linear combination of these vectors with the relative contributions determined by constructing quadratic models of S and χ^2. The χ^2 parameter and its gradient and curvature are usually calculated using fast Fourier transforms and no full $N_\mathrm{f} \times N_\mathrm{f}$ matrix operations are involved.

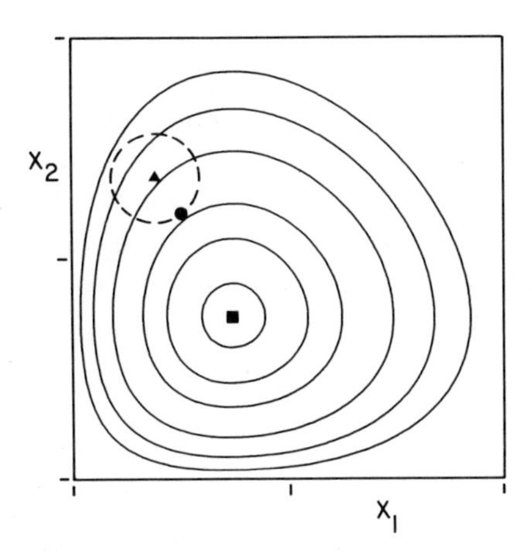

Fig. 3.3. Contour plot of the entropy S of (3.5) (solid lines) and $\chi^2 = 2N_t$ (3.6) (dashed line) for a spectrum consisting of two intensities x_1 and x_2. The square represents the unconstrained entropy maximum ($x_1 = x_2 = b/\mathrm{e}$); the triangle, an exact fit to the data ($\chi^2 = 0$) and the circle, the constrained entropy maximum.

The default parameter b serves to incorporate prior knowledge about the spectrum baseline, as follows. In the absence of data, the maximum (unconstrained) entropy is obtained with a completely uniform spectrum, having all the x_n equal to b/e. The χ^2 constraint tends to introduce structure at frequencies where there is evidence for a signal in the data, but has little effect at other frequencies. Setting b equal to a number much smaller than the expected peak intensities therefore ensures that the baseline has low intensity. An alternative definition of entropy has $\sum_n x_n$ in place of b, so that the spectral intensities are normalized. This has the disadvantage of pulling the baseline up to an unrealistically high level.

To summarize: in its simplest form, maximum entropy processing of NMR data leaves only two parameters at the disposal of the operator, N_f and b. N_f can have any value greater than or equal to $2N_t$, the number of samples in the free induction decay. A large value of N_f gives a finely digitized spectrum, at the expense of increased processing time. If a fast Fourier transform algorithm is used to transform trial spectra into trial data, N_f should be an integral power of two. The noise level σ may be calculated from the tail of the free induction decay or from a separate signal recorded under identical conditions but with no radiofrequency pulses. No prior knowledge or prejudice concerning the number of resonances in the spectrum or their lineshapes is required. However, if the lineshapes *are* known, then this information can, and should, be included in the transformation $\mathbf{R}$ (3.7) and will give spectra of higher quality (see Section 3.5.2).

3.5 Problems

Those who use NMR spectrometers are fortunate in that the signals they record are, usually, remarkably free from distortions and defects. NMR is less in need of fancy data processing than many other experimental techniques, some of which are discussed elsewhere in this book. Nevertheless, NMR signals are seldom perfect: in the following paragraphs, four common problems are discussed.

3.5.1 Sensitivity

Perhaps the major disadvantage of NMR is its low sensitivity. Unlike most other forms of spectroscopy, NMR probes energy differences that are tiny compared to thermal energies. Population differences between spin states are therefore small, leading to feeble NMR intensities and poor signal-to-noise ratios, which restrict the accuracy to which intensities, frequencies and relaxation times can be measured, and limit possible applications. It is therefore essential to consider carefully ways of enhancing signal strength and attenuating noise. While there are many experimental methods of improving sensitivity (Ernst *et al.* 1987, p. 2), the data processing options are much more limited. But first, let us be clear about the meaning of the terms sensitivity and signal-to-noise ratio (Freeman 1987, p. 122).

Sensitivity relates to the ability to detect weak resonances, that is to distinguish them from noise. A technique can be said to enhance sensitivity if it would allow the detection of a previously unobservable signal. The signal-to-noise ratio is a number determined from the spectrum, defined as the peak signal height divided by twice the root-mean-square noise level, calculated from a region of the spectrum devoid of signal (the baseline). Usually, improvements in sensitivity and signal-to-noise ratio go hand in hand and the terms are used interchangeably. However, this need not be so when the spectrum is obtained by non-linear methods, as the following example demonstrates. Fig. 3.4a shows a conventional spectrum, obtained from a synthesized free induction decay by Fourier transformation with no sensitivity or resolution enhancement. The spectrum in Fig. 3.4b is calculated by subtracting a constant level (indicated by the arrows in Fig. 3.4a) from the Fourier transform spectrum and setting the negative intensities equal to zero. This 'scissors' operation removes almost all of the noise on the baseline and so gives an enormous improvement in signal-to-noise ratio. The sensitivity, however, is unchanged: had there been a weak resonance lost in the noise in Fig. 3.4a, it would have been excised along with the noise.

To improve sensitivity by data processing alone, one must exploit the properties of noise that distinguish it from genuine signals. In the frequency domain, signals are smooth and (usually) sharp while noise is random and

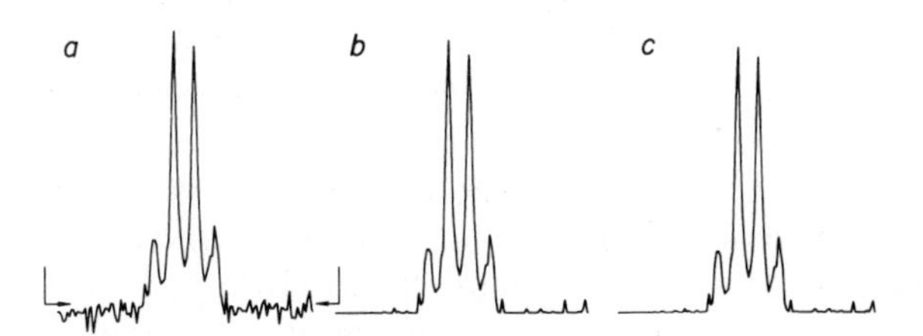

Fig. 3.4. Spectra calculated from a synthesized free induction decay comprising four exponentially damped sinusoids with equally spaced frequencies, and intensities in the ratio 1 : 3 : 3 : 1. (*a*) Real part of the Fourier transform spectrum. (*b*) 'Scissors' spectrum (see text). (*c*) Maximum entropy reconstruction.

has a broad frequency spectrum. One may discriminate between the two by some form of digital filtering, as mentioned earlier. This may be done in the frequency domain by convolving the spectrum with a smoothing (bell-shaped) function or, equivalently and more conveniently, by multiplying the free induction decay by a weighting function (for example, a decaying exponential) prior to Fourier transformation. This process suppresses the noise in the tail of the free induction decay but also damps the signals, and so broadens lines in the spectrum. The optimum sensitivity enhancement function is the one that exactly matches the decay of the signal: this 'matched filter' gives the maximum signal-to-noise ratio and, for Lorentzian lines (that is, exponential relaxation), doubles the line width (Ernst 1966). No sustainable case has yet been made for sensitivity enhancement in excess of that yielded by matched filtering.

This is not to say, however, that no such claims have been made. An article in *The Times* of London (Redfearn 1984), based loosely on a letter to *Nature* (Sibisi *et al.* 1984), went so far as to suggest that maximum entropy would improve sensitivity to the extent that carbon-13 NMR spectra could, in future, be performed routinely on living creatures without isotopic enrichment. Similar, but less extreme, statements can be found in several papers in the NMR literature (Sibisi 1983; Sibisi *et al.* 1984; Ni *et al.* 1986; Mazzeo and Levy 1986). The basis of such claims stems from the rather seductive philosophy of the maximum entropy method, as outlined above, and from confusion between the terms sensitivity and signal-to-noise ratio.

The latter point is illustrated by Fig. 3.4*c*, which shows the maximum entropy reconstruction of the synthetic free induction decay used to produce Fig. 3.4*a*,*b*. Note the excellent attenuation of baseline noise. Closer inspection, however, reveals the failure of maximum entropy to remove noise at frequencies corresponding to genuine signals. The distortions of the peak heights from 1 : 3 : 3 : 1, due to the underlying noise, are clearly present in both maximum entropy and Fourier transform spectra. Thus the only achievement of maximum entropy processing has been to clean up the uninteresting regions of the spectrum where there are no resonances. Indeed, there is a striking similarity between this reconstruction and the scissors spectrum (a relationship predicted by Hoch *et al.* (1990)). Clearly, in this case, maximum entropy has not improved sensitivity. But then, we have not used any information about the lineshape and so should not expect to get even the sensitivity improvement afforded by matched filtering of the conventional Fourier transform spectrum. This point is discussed further in Section 3.5.2.

3.5.2 Resolution

Compared to most other forms of spectroscopy, NMR enjoys staggeringly impressive resolution: lines as close as 1 part in 10^9 may routinely be resolved for small molecules in non-viscous solvents. Nevertheless, the small spread of resonance frequencies (typically 1 part in 10^5 for protons) can cause substantial crowding especially in the spectra of large molecules with many, relatively broad lines. Furthermore, long range spin–spin couplings (useful in spectral assignment and structure determination) are often similar in size to line widths and so difficult to resolve.

Resolution may be improved by multiplying the free induction decay prior to Fourier transformation by a weighting function that initially rises and then falls to a small value at the end of the data (to avoid introducing a step-function discontinuity, see Section 3.5.3). This procedure removes some of the damping of the free induction decay, and so narrows the lines in the spectrum; at the same time, it amplifies the noise and so degrades sensitivity. Any attempt at digital filtering must always be a compromise between sensitivity and resolution. Put another way, 'you can't get something for nothing', a principle that applies equally to linear and non-linear data processing.

Maximum entropy spectra with enhanced resolution may be obtained (Laue *et al.* 1986) by introducing a weighting function into the transformation $\mathbf{R}$ (3.7). The idea is that trial spectra containing sharp lines are transformed into slowly decaying trial data, which in turn are weighted, so that their decay parallels that of the actual data.

To judge the ability of maximum entropy to enhance resolution, we must consider a spectrum consisting of at least two lines. This is because line width and resolution, like signal-to-noise ratio and sensitivity, are not necessarily related for spectra obtained by non-linear methods. The line width of a resonance is a number, for example the full width at half maximum height. Resolution is more subtle and refers to the ability to separate overlapping resonances. This distinction may be illustrated by considering non-linear amplification of a doublet resonance in which the splitting is not

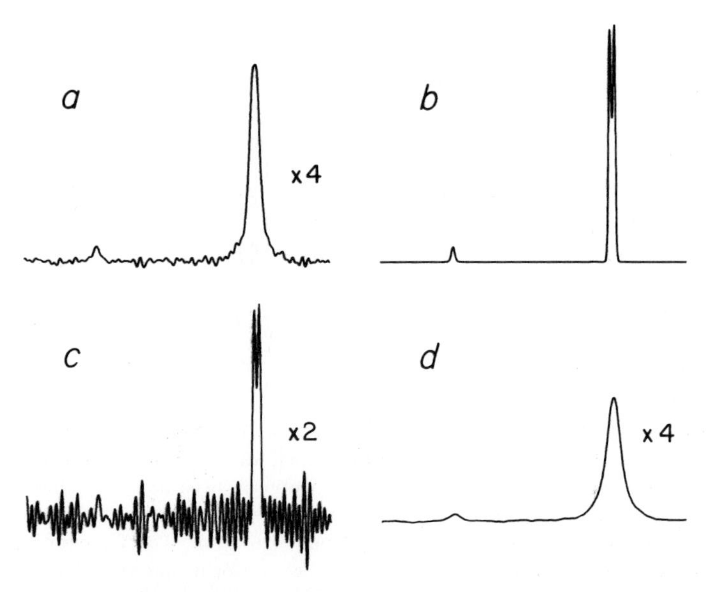

Fig. 3.5. Spectra obtained from a synthesized free induction decay. (*a*) Real part of the Fourier transform spectrum with no resolution enhancement. (*b*) Maximum entropy reconstruction with resolution enhancement. (*c*) Real part of the Fourier transform spectrum with resolution enhancement. (*d*) Real part of the Fourier transform spectrum with matched filtering. Note the vertical expansions of *a*, *c* and *d*.

resolved. We could imagine applying an amplification so non-linear as to narrow the line down to almost nothing. This operation will not cause the splitting to become apparent and so does not affect resolution.

As an example of resolution enhancement by maximum entropy (Jones and Hore, in press), we take a synthesized free induction decay comprising a strong doublet and a weak singlet, plus noise (Fig. 3.5). The doublet splitting is chosen such that the two components accumulate a phase difference of 2π by the end of the free induction decay. Fig. 3.5 shows four spectra calculated from this free induction decay. Fig. 3.5*a* is the conventional spectrum: the doublet is unresolved and the singlet just visible above the noise. Fig. 3.5*b* is the maximum entropy reconstruction, with resolution enhancement, using as weighting function an exponential matched to the decay of the data. The doublet splitting is now clear and the noise on the baseline essentially absent. Fig. 3.5*c* is the Fourier transform spectrum with resolution enhancement. The weighting function was chosen so as to convert the lineshape to Gaussian form with line-width similar to that in the maximum entropy reconstruction. In this spectrum the singlet has all but disappeared into the noise. For comparison, Fig. 3.5*d* shows the Fourier

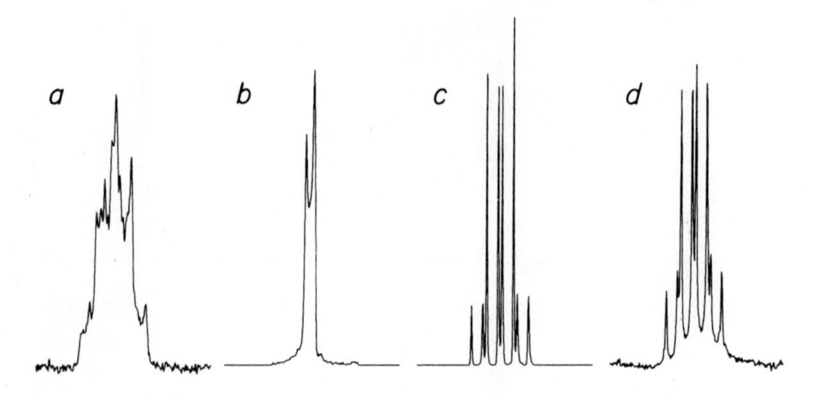

Fig. 3.6. Resolution enhancement of data recorded in an inhomogeneous magnetic field. (*a*) Part of the conventional spectrum with no resolution enhancement. (*b*) Another section from the same spectrum, used as a reference lineshape for deconvolution. (*c*) Maximum entropy reconstruction using the inverse Fourier transform of (*b*) as a resolution enhancement function. (*d*) Conventional spectrum calculated from data recorded in a more uniform magnetic field. The four spectra are arbitrarily scaled. (Courtesy of D. S. Grainger.)

transform spectrum with matched filtering for maximum sensitivity.

Fig. 3.5*b* reveals clearly, in a single spectrum, the existence of a weak singlet and a strong doublet with no evidence for further resonances. To obtain this information as convincingly by conventional methods, one would require at least two Fourier transform spectra—one with resolution enhancement to resolve the doublet splitting and another, preferably with sensitivity enhancement, to locate the singlet. One can envisage circumstances in which it will be advantageous to view a single spectrum containing all the information in the data, rather than several, displaying different aspects of that information.

Just whether this constitutes an improvement in sensitivity or resolution is doubtful. In the author's experience, it has proved impossible to reveal a weak resonance by maximum entropy that could not be located in the matched-filtered conventional spectrum. Similarly, I have not yet been able to resolve a splitting using maximum entropy that could not also be seen in a resolution-enhanced Fourier transform spectrum.

Finally, we present an example using genuine data. Fig. 3.6*a* shows one multiplet taken from a spectrum obtained by Fourier transformation without resolution enhancement. The homogeneity of the static magnetic field is poor, giving broad distorted lineshapes, such that the structure of the multiplet is not apparent even with conventional resolution enhancement (not shown). Just how bad the lineshape is may be appreciated from

Fig. 3.6*b*, which shows another section of the same spectrum. This resonance, which should be a sharp singlet, has been split into an asymmetric doublet by the inhomogeneity of the field. Since this distortion will be the same for all lines in the spectrum (all molecules experience the same non-uniform field), we may use this resonance to deconvolve the multiplet of interest. To be specific, the reference lineshape (Fig. 3.6*b*) is transformed back into the time domain and is used as the weighting function in a maximum entropy reconstruction. The resulting spectrum (Fig. 3.6*c*) shows clearly that the multiplet is a doublet of quartets. The integrated intensities of the eight lines are in the approximate ratios $1:1:3:3:3:3:1:1$. This interpretation is confirmed by Fig. 3.6*d*, the Fourier transform of a free induction decay obtained with a much more uniform field.

The moral, of course, is that it is preferable to record data in a uniform magnetic field: however, if this is not possible, then maximum entropy is a good way of 'making the best of a bad job'.

This procedure exemplifies one aspect of the maximum entropy philosophy: all known instrumental distortions should be incorporated into the transformation from the frequency domain to the time domain in order that the trial spectra are the sort that might be obtained from a perfect spectrometer operating under ideal conditions. Failure to do so will result in low quality spectra, with instrumental distortions present (Davies *et al.* 1988). Note that the close similarity between the maximum entropy reconstruction and a scissored version of the conventional spectrum is not observed when a resolution enhancement function is included in $\mathbf{R}$ (nor when the data are truncated: see Section 3.5.3).

Further examples of resolution enhancement of NMR spectra are discussed in Chapter 4.

3.5.3 Truncation

In some NMR experiments, it is desirable to stop data acquisition before the free induction decay has decayed to a negligible level. Such 'truncated' signals pose a problem if the spectrum is to be determined by Fourier transformation. There are two approaches (Fig. 3.7). One may assume the missing data points are all equal to zero (a clearly unfounded, if convenient, fiction) and then Fourier transform. This procedure produces a spectrum with 'sinc-wiggles'; that is, one gets the desired spectrum (Fig. 3.7*b*) convolved with a sinc function ($\text{sinc}(x) = \sin(\pi x)/\pi x$), Fig. 3.7*d*. The side-lobes of this lineshape may obscure genuine signals or be mistaken for such. Alternatively one may apodize the experimental signal by multiplying it by a sensitivity enhancement function (usually exponential) prior to zero-filling and Fourier transformation. This weighting forces the free induction decay to die away smoothly, avoiding the sinc-wiggles but broadening the lines and reducing resolution, Fig. 3.7*f*. In the first case data

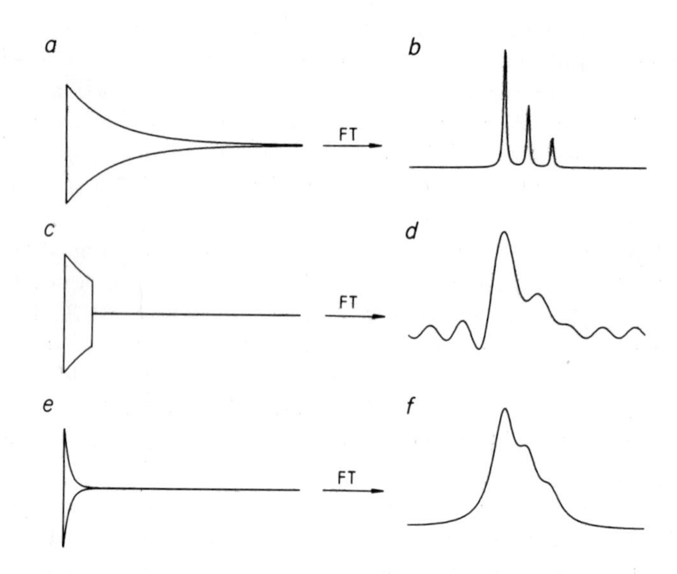

Fig. 3.7. Schematic free induction decays (*a*,*c*,*e*) and their Fourier transforms (*b*,*d*,*f*). Only the envelopes of the time domain signals and the real parts of the spectra are shown. (*a*,*b*) Complete signal. (*c*,*d*) Truncated signal with zero-filling. (*e*,*f*) Truncated signal, apodized by exponential weighting and zero-filled. FT denotes Fourier transformation.

are invented and in the second, good information is thrown away: both are clearly unsatisfactory. While these problems can often be avoided simply by extending the acquisition time, limited computer memory or constraints on the overall duration of the experiment can sometimes make truncated responses unavoidable (see Section 3.6).

The maximum entropy method seeks to circumvent these problems by proceeding in the opposite direction, namely from the frequency domain to the time domain (Hoch 1985; Hore 1985; Laue *et al.* 1986). As indicated in Fig. 3.8, trial spectra with (one hopes) undistorted lineshapes are inverse Fourier transformed to give trial data with no truncation. The available experimental data are then compared with the corresponding portion of the trial free induction decays by calculating χ^2.

Sinc-wiggles are suppressed in two ways. First, since they decrease the entropy (by increasing the information content) without affecting χ^2, there is no incentive to introduce them into the trial spectra. Second, the trial spectra must have strictly positive intensities (because of the logarithmic term in (3.5)): this excludes the negative lobes of the sinc function. Positivity also has the effect of suppressing the positive lobes which would otherwise produce a drastic change in χ^2, for which there would be no

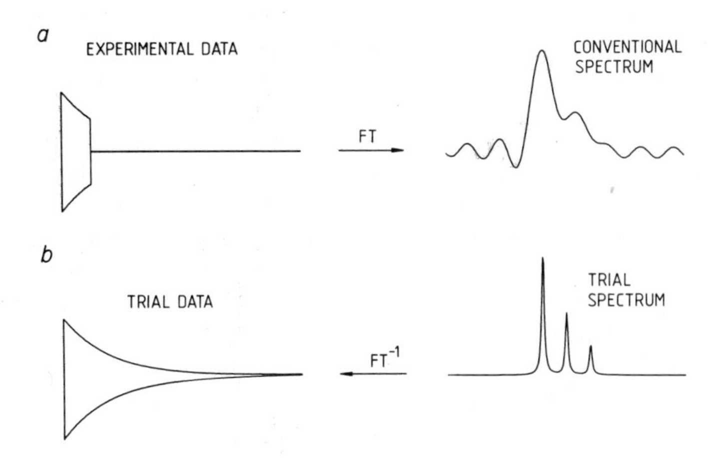

Fig. 3.8. Treatment of truncated signals by (a) conventional and (b) maximum entropy methods. The available experimental data are compared with the corresponding portion of the trial data. Only the envelopes of the time domain signals and the real part of the conventional spectrum are shown. FT and FT^{-1} denote Fourier and inverse Fourier transformation, respectively.

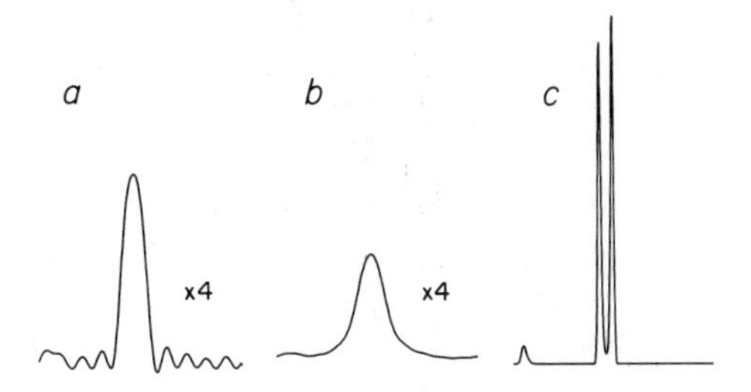

Fig. 3.9. Spectra obtained from a synthesized free induction decay. (a) Real part of the Fourier transform spectrum. (b) Real part of the Fourier transform spectrum with strong exponential apodization. (c) Maximum entropy reconstruction with resolution enhancement. Note the vertical expansions of a and b.

evidence in the data.

An illustration (Jones and Hore 1990) of the way in which maximum entropy copes with truncated data is shown in Fig. 3.9 for a synthetic free induction decay, comprising a strong doublet, a weak singlet and noise. The signal is severely truncated and the doublet splitting is such that the two components accumulate a phase difference of $11\pi/8$ by the end of the data. Fig. 3.9a shows the conventional spectrum, with zero-filling, but no weighting: the doublet is not resolved and the singlet cannot confidently be distinguished from the sinc-wiggles and the noise. Fig. 3.9*b* is the same spectrum with strong exponential weighting to suppress sinc-wiggles: the lines are much broader and the singlet can just be discerned. The maximum entropy reconstruction, using an exponential weighting function for resolution enhancement (see Section 3.5.2), however, shows clear evidence for both the singlet and the doublet (Fig. 3.9*c*). Although the doublet components have different heights and widths (because of the noise), their integrated intensities are very similar. To resolve the doublet splitting using conventional filtering, a very severe weighting function is required, greatly amplifying the sinc-wiggles.

Once again, it seems that the maximum entropy reconstruction reveals spectral features that cannot be seen in a *single* conventional spectrum. Further examples of maximum entropy processing of truncated signals will be discussed in Section 3.6.

3.5.4 Dispersion lineshapes

The ideal to which most NMR experiments aspire is the absorption-mode lineshape. Dispersion-mode peaks (Fig. 3.2 and Eqn 3.4), with their broad wings and antisymmetric shape degrade both sensitivity and resolution, and hinder accurate determination of resonance frequencies and line intensities. For these reasons, NMR spectroscopists often go to great lengths when designing their experiments to ensure that the real part of the Fourier transform spectrum contains only absorption lineshapes, even if this means sacrificing sensitivity. At first sight, the dispersion lineshape would seem to be a fundamental and inescapable feature of NMR. In reality, it is simply an artefact of the way NMR data are normally recorded and processed.

In most NMR experiments, one detects the magnetization excited by a radiofrequency pulse or sequence of pulses. The signal is inherently 'one-sided' ($t > 0$), with no data recorded during the time prior to excitation ($t < 0$). To understand how dispersion lineshapes arise, consider the hypothetical 'double-sided' NMR signal with phase angle ϕ, recorded for positive *and* negative times:

$$f(t) = \exp(\mathrm{i}\Omega t)\exp(-|t|/T_2)\exp(\mathrm{i}\phi). \tag{3.8}$$

Note that the envelope of $f(t)$, $\exp(-|t|/T_2)$, is symmetrical in t. Fourier transformation of this signal gives the spectrum

$$F(\omega) = 2A(\omega)\exp(\mathrm{i}\phi), \tag{3.9}$$

where $A(\omega)$ is the absorption lineshape (3.4). $F(\omega)$ has no dispersion component because the Fourier transform of an oscillating function with a symmetrical envelope gives a spectrum with symmetrical lineshapes.

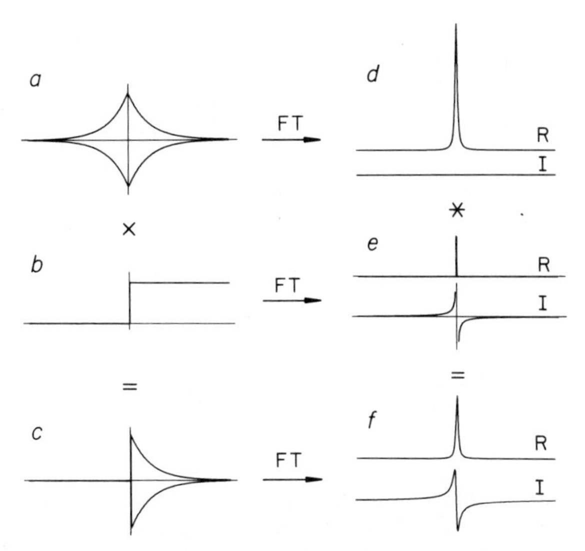

Fig. 3.10. The origin of dispersion lineshapes. (a) Double-sided free induction decay $f(t)$. (b) Step function $h(t)$. (c) One-sided free induction decay $s(t)$. d, e and f are Fourier transforms of, respectively, a, b and c. Note that $s(t) = f(t) \times h(t)$ and $S(\omega) = F(\omega) * H(\omega)$. Only the envelopes of the time domain signals are shown. R, I and FT denote real and imaginary parts, and Fourier transformation, respectively.

A normal, one-sided, free induction decay $s(t)$ would be similar to $f(t)$ for $t > 0$ but undetermined for $t < 0$. Assuming that these missing data are equal to zero, $s(t)$ is equal to the product of $f(t)$ and the step function $h(t)$ defined by

$$h(t) = \begin{cases} 0, & \text{if } t < 0; \\ 1, & \text{if } t > 0. \end{cases} \tag{3.10}$$

Put another way, the Fourier transform spectrum $S(\omega)$ corresponding to $s(t)$ is the convolution of $F(\omega)$ with $H(\omega)$, where $H(\omega)$ is the Fourier transform of $h(t)$:

$$H(\omega) = \pi\delta(\omega) - \mathrm{i}/\omega, \tag{3.11}$$

$$S(\omega) = F(\omega) * H(\omega) = \big(A(\omega) + \mathrm{i}D(\omega)\big)\exp(\mathrm{i}\phi). \tag{3.12}$$

The two equivalent operations, multiplication in the time domain and convolution in the frequency domain, are shown in Fig. 3.10. It can be seen that the dispersion lineshape arises from convolution of $A(\omega)$ with i/ω, the broad, antisymmetric part of $H(\omega)$. More fundamentally, $D(\omega)$ is produced by Fourier transforming one-sided data, with the assumption that the missing data for negative times are equal to zero.

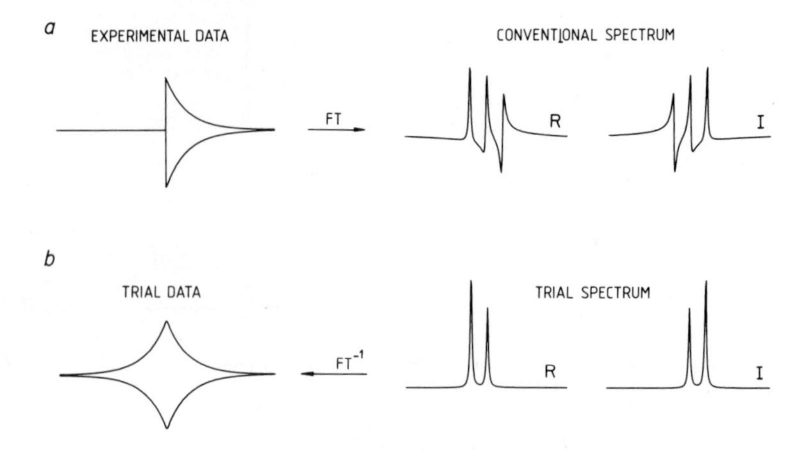

Fig. 3.11. Treatment of one-sided signals by (*a*) conventional and (*b*) maximum entropy methods. As in Fig. 3.8, the available experimental data are compared with the corresponding portion of the trial data. Only the envelopes of the time domain signals are shown. R, I, FT and FT^{-1} denote real and imaginary parts, Fourier and inverse Fourier transformation, respectively.

In many cases the dispersion component in $S(\omega)$ presents no difficulties at all. If the experiment is such that all resonances have one of two phases, 0 or π, the dispersion components are confined to the imaginary part of the spectrum which may be discarded. However, when resonances with phases other than 0 and π are present, dispersion-mode peaks will inevitably appear in the real part of the spectrum.

Maximum entropy processing of one-sided NMR data (Fig. 3.11) is essentially identical to the treatment of truncated signals (Fig. 3.8) except that one needs a complex trial spectrum (Daniell and Hore 1989). The criterion of agreement between trial and experimental data (χ^2) is again calculated by summing over the available experimental points ($t > 0$). Just as maximum entropy reconstructions of truncated data are largely free from sinc-wiggles, so one can anticipate the absence of dispersion lineshapes in reconstructions of one-sided data.

An example of the use of maximum entropy to separate resonances of different phase will be found in Section 3.7.

3.6 Examples

The following paragraphs describe two experiments that illustrate the use of the maximum entropy method in NMR. Both belong to a large class of

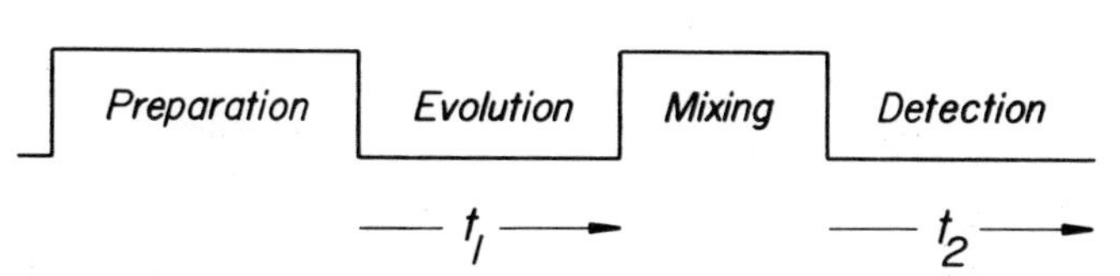

Fig. 3.12. The four constituent parts of a radiofrequency pulse sequence for two-dimensional NMR.

techniques known as two-dimensional NMR (Ernst *et al.* 1987; Freeman 1987, p. 291). The molecules on which these experiments were performed were chosen, not for their intrinsic chemical interest, but so as to bring out most clearly the differences between maximum entropy and Fourier transform spectra.

The radiofrequency pulse sequences necessary to obtain two-dimensional NMR data consist, in general, of four periods: preparation, evolution, mixing and detection (Fig. 3.12). Initially, the spins are prepared in a well-defined coherent state by one or more radiofrequency pulses. There follows a period of coherent evolution during which the spins precess under the influence of the static magnetic field, spin–spin couplings and any applied radiofrequency fields. After a time t_1, the coherences present during the evolution period are transformed by means of one or more pulses (the mixing period) into xy magnetization which is detected as a function of time t_2. To sample the evolution during t_1, a series of free induction decays is recorded with different, usually equally spaced, values of t_1 to obtain a matrix of data $s(t_1, t_2)$. Ignoring relaxation and phase factors, $s(t_1, t_2)$ is a sum of components of the form

$$\exp(\mathrm{i}\Omega_1 t_1)\exp(\mathrm{i}\Omega_2 t_2) \qquad \text{or} \qquad \cos(\Omega_1 t_1)\exp(\mathrm{i}\Omega_2 t_2), \tag{3.13}$$

that is, a collection of conventional free induction decays $\exp(\mathrm{i}\Omega_2 t_2)$, phase- or amplitude-modulated as a function of t_1 at frequency Ω_1. Conventional processing involves Fourier transformation of $s(t_1, t_2)$ with respect to t_1 and t_2 to obtain a two-dimensional spectrum as a function of two frequency variables ω_1 and ω_2. For every signal of the form of (3.13), $S(\omega_1, \omega_2)$ would have a peak at frequency coordinates $\omega_1 = \Omega_1$ and $\omega_2 = \Omega_2$.

The sort of information contained in the two-dimensional spectrum is determined by what happens during the four periods, and is under the control of the spectroscopist. For example, one can arrange the pulse sequence so that the occurrence of a peak at frequency coordinates (Ω_1, Ω_2) indicates a correlation (for example, a spin–spin coupling: see Section 3.6.2) between a spin with resonance frequency Ω_1 and another with resonance frequency Ω_2 (Aue *et al.* 1976). Experiments such as these have been enormously influential over the last ten years in extending the range of systems

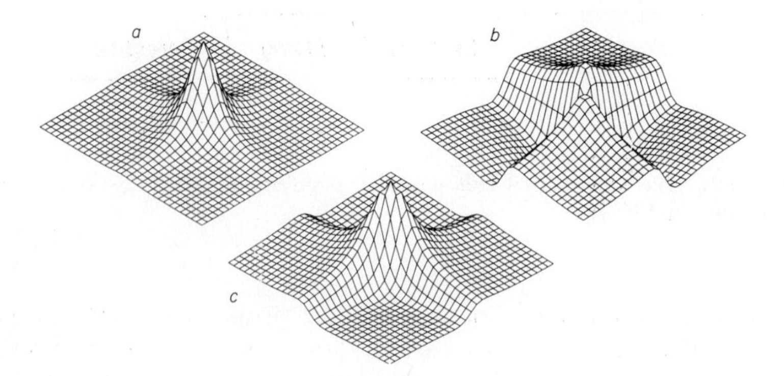

Fig. 3.13. Two dimensional NMR lineshapes. (*a*) Pure absorption-mode. (*b*) Pure dispersion-mode. (*c*) The phase-twist lineshape.

amenable to study by NMR. The improvement in resolution obtained by spreading resonances out into a second frequency dimension has allowed, for example, the complete three-dimensional structures of small proteins to be determined (Wüthrich 1986).

Finally a word on two-dimensional NMR lineshapes: these are usually products of absorption or dispersion lineshapes (3.4) in the two dimensions or linear combinations thereof. Three commonly occuring lineshapes (Fig. 3.13) are pure absorption $A(\omega_1)A(\omega_2)$, pure dispersion $D(\omega_1)D(\omega_2)$ and the phase-twist, a superposition of pure absorption and pure dispersion, $A(\omega_1)A(\omega_2)-D(\omega_1)D(\omega_2)$. The undesirable broad wings and negative regions of both the pure dispersion and the phase-twist are plainly visible.

3.6.1 Rotating frame imaging

The aim of most NMR experiments is to get a uniform response from all parts of a homogeneous sample. This is achieved by making both the static magnetic field and the radiofrequency field as spatially uniform as possible. However, there are a number of techniques which use non-uniform magnetic fields to obtain NMR spectra from different regions of heterogeneous samples, that is with spatial as well as frequency resolution (Morris 1986). A simple example is the two-dimensional experiment known as rotating frame imaging (Hoult 1979; Cox and Styles 1980; Styles *et al.* 1985), the pulse sequence for which is shown in Fig. 3.14*a*. (The term 'rotating frame' refers to a simple, pictorial description of NMR, based on a coordinate system rotating around the z-axis at the frequency of the electromagnetic field (Bloch 1956). In the remainder of this section, references to resonance frequencies will be in the rotating frame, that is relative to the frequency of the electromagnetic field.)

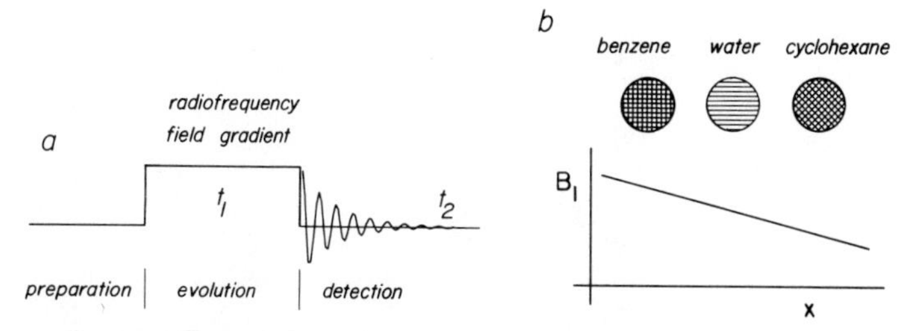

Fig. 3.14. (a) Pulse sequence for rotating frame imaging. (b) Sample consisting of three liquid filled glass bulbs in a linear radiofrequency field gradient $B_1(x)$.

In this experiment, the mixing period is absent and the preparation period consists of waiting for the spins to reach equilibrium. During the evolution period t_1, the spins precess in a linear radiofrequency field gradient directed along the x axis, with frequency

$$\Omega_1 = \gamma\big(B_1(x)^2 + \Delta B^2\big)^{\frac{1}{2}}. \tag{3.14}$$

$B_1(x)$ is the magnitude of the radiofrequency field at position x, and ΔB is the offset from resonance ($\gamma\Delta B$ is the resonance frequency in the rotating frame: ΔB is determined by chemical shifts and spin–spin couplings). The quantity γ is the gyromagnetic ratio ($2.675 \times 10^8\,\mathrm{T}^{-1}\mathrm{s}^{-1}$ for protons). At the end of t_1, the radiofrequency field is switched off and the spins undergo free precession during the detection period t_2 in a homogeneous static field at frequencies determined by their offset from resonance: $\Omega_2 = \gamma\Delta B$. The key to the successful working of the experiment is to choose the radiofrequency field strength such that it is much larger than the resonance offset for all the spins, irrespective of their resonance frequency or position in the sample: $B_1(x) \gg |\Delta B|$. Under these conditions, $\Omega_1 = \gamma B_1(x)$ and the two-dimensional data will consist of components of the form

$$\cos\big(\gamma B_1(x)t_1\big)\exp(\mathrm{i}\gamma\Delta B t_2) \tag{3.15}$$

with amplitude modulation as a function of t_1. The frequency in the ω_1 dimension should reflect only the position of the spins in the sample, while the ω_2 frequency should be the resonance offset. Thus spatial and frequency information are separated into the two dimensions.

As a simple example, suppose we have a sample comprising three glass bulbs containing benzene, water and cyclohexane (molecules with no spin–spin couplings and different chemical shifts) aligned along the x axis. We arrange the radiofrequency field gradient to be linear in x with the benzene experiencing the largest field and cyclohexane the smallest (Fig. 3.14b).

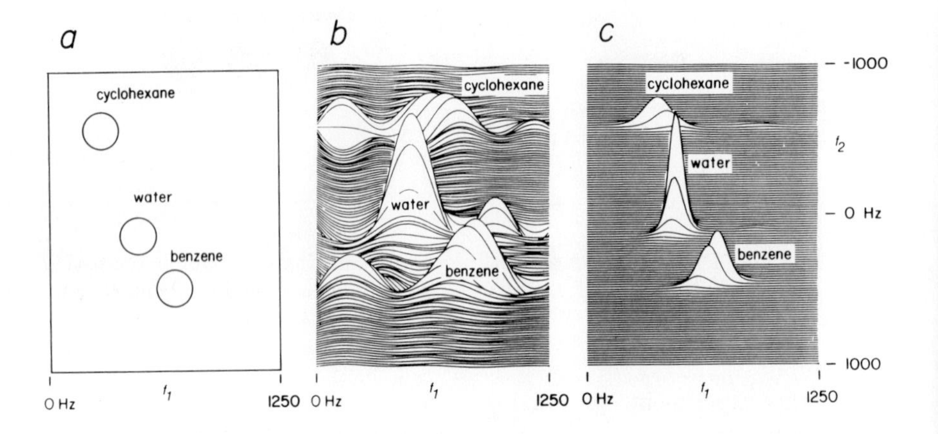

Fig. 3.15. Rotating frame images of the sample shown in Fig. 3.14*b*. (*a*) Schematic spectrum. (*b*) Fourier transform spectrum. (*c*) Maximum entropy reconstruction. Note that the f_1 axis corresponds to the position in the sample, with the radiofrequency field gradient decreasing from right to left.

The three samples have distinct positions in space and different chemical shifts, so that three peaks should appear in the two-dimensional spectrum, as shown schematically in Fig. 3.15*a*.

Problems arise when the spins resonate at an offset ΔB from the radiofrequency carrier which is not very much smaller than the local radiofrequency field strength B_1. When such off-resonance effects are significant, the data no longer have the simple form in (3.15). The most serious consequence is that Ω_1, the precession frequency of a nucleus during t_1, no longer depends solely on its position in the sample but also, misleadingly, on the resonance offset (3.14). Other distortions include unwanted phase shifts and the appearance of extraneous signals along the $\omega_1 = 0$ axis (see below).

The severity of these effects may be judged from experimental spectra (Hore and Daniell 1986) obtained from the sample described above (Fig. 3.14*b*). The conventional spectrum (calculated by double Fourier transformation of the data) is shown in Fig. 3.15*b*. In addition to sinc-wiggles caused by truncation (mild in t_2 but severe in t_1), three distortions, arising from off-resonance effects, are noticeable. First, there are axial peaks with pure dispersion lineshapes close to $\omega_1 = 0$ at the ω_2 frequencies corresponding to the chemical shifts of the three molecules. Second, the genuine peaks have mixed absorption and dispersion lineshapes. Third, and most disturbing, the peaks are shifted in the ω_1 dimension because of the dependence of Ω_1 on ΔB (3.14). This effect is most severe for cyclohexane which experiences a small radiofrequency field and has a large resonance offset from the transmitter. The ω_1 frequencies in Fig. 3.15*b* suggest that the cyclohexane lies between water and benzene, when in fact water is in the middle.

The blame for these distortions and artefacts may be laid at the feet of the Fourier transform. When the off-resonance effects are significant, the Fourier transform simply is not the correct method of obtaining a spectrum. What is needed is a technique that recognizes the dependence of the Ω_1 frequencies on ΔB as well as on $B_1(x)$: maximum entropy is one such method. It circumvents these difficulties by proceeding, as usual, from frequency domain to time domain and by using the appropriate transformation calculated from the geometry of precession in the rotating frame, using the Bloch equations (Bloch 1956).

The maximum entropy spectrum (Fig. 3.15*c*) is largely free from distortion. The sinc-wiggles, axial peaks, and dispersive contributions can barely be seen and the peaks are correctly ordered in the distance dimension. This spectrum should be compared with Fig. 3.15*a*.

The principal use of this type of experiment is to record spatially localized, *in vivo* phosphorus-31 NMR spectra of animals and human beings without the need for surgery. In practice, off-resonance effects in rotating frame imaging are often inevitable because safety considerations preclude the use of high transmitter powers (radiofrequency radiation causes tissue heating). Truncation in t_1 is also necessary to limit the time patients must lie motionless inside an NMR magnet.

3.6.2 Chemical shift correlation

One of the simplest and most useful two-dimensional NMR experiments is chemical shift correlation, also known as COSY (*CO*rrelated *S*pectroscop*Y*) (Aue *et al.* 1976). The pulse sequence consists of two 90° pulses, separated by the evolution period t_1, and followed by the detection period t_2 (Fig. 3.16). (As its name suggests, a 90° pulse rotates the magnetization of the sample through 90°.) The first pulse (preparation) excites the spins, rotating them from the z axis into the xy plane after which they precess coherently at their resonance frequencies for a time t_1. The second pulse (mixing) transfers this coherence between pairs of spin–spin coupled nuclei, and is followed by a second period of free precession during which the signal is detected.

Consider a pair of coupled nuclei, A and B, with resonance frequencies Ω_A and Ω_B. The two-dimensional COSY spectrum $S(\omega_1, \omega_2)$ consists of two types of peak, diagonal and cross. Diagonal peaks are centred at the same frequency in the two dimensions, (Ω_A, Ω_A) and (Ω_B, Ω_B), and are due to magnetization that precesses at the same frequency during t_1

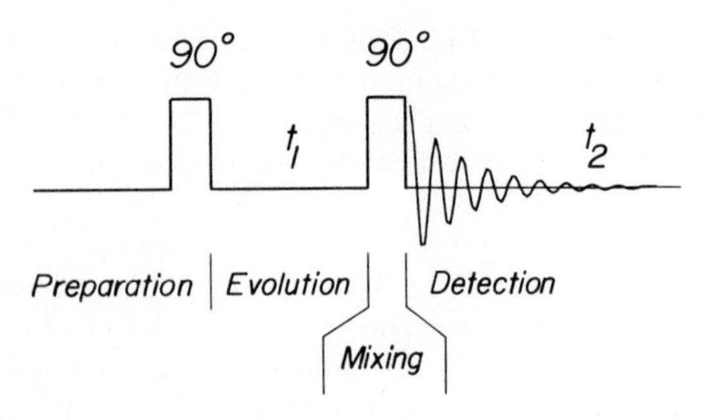

Fig. 3.16. Radiofrequency pulse sequence for chemical shift correlation.

and t_2. These responses contain no more information than the normal one-dimensional spectrum. Cross peaks are centred at different frequencies in the two dimensions, (Ω_A, Ω_B) and (Ω_B, Ω_A), and arise from magnetization which precesses at one frequency during t_2 and is modulated at another frequency as a function of t_1. The appearance of such peaks indicates that nuclei A and B are spin–spin coupled and must therefore be in the same molecule, separated by a small number of chemical bonds. Information of this sort is crucial in assigning NMR spectra (the process of deciding which nucleus in the molecule corresponds to which resonance in the spectrum).

To illustrate the technique, we consider two small portions of the ^{1}H COSY spectrum of 2,3-dibromopropionic acid, a molecule containing three protons, each of which is coupled to the other two. Fig. 3.17*a,b* shows, schematically, parts of two cross-sections running through the cross-peaks centred at $(\omega_1, \omega_2) = (\Omega_X, \Omega_A)$ and (Ω_X, Ω_M) respectively (A, M and X label the three protons: see Fig. 3.17). Both traces are parallel to the ω_1 axis, and show the resonance of spin X, split by spin–spin coupling with A and M. In these cross-sections the couplings may be classified as either active (the coupling that is responsible for the cross-peak) or passive (all other couplings). The splittings due to active couplings (J_{AX} in Fig. 3.17*a* and J_{MX} in Fig. 3.17*b*) appear in anti-phase, that is, with the components of the doublet 180° out of phase. The passive splittings (J_{MX} in Fig. 3.17*a* and J_{AX} in Fig. 3.17*b*) are in phase, that is, both components of the doublet have the same phase.

Before proceeding, let us pause to consider the conditions necessary to observe these cross-peaks. Because of the anti-phase nature of the active splittings, it is essential that they are moderately well resolved, otherwise the anti-phase components cancel and the cross-peaks disappear into the noise. The fundamental limit on resolution is the duration of the measurement, which should exceed the inverse of the smallest splitting to be resolved. This puts a lower limit on the number of t_1 values that must be

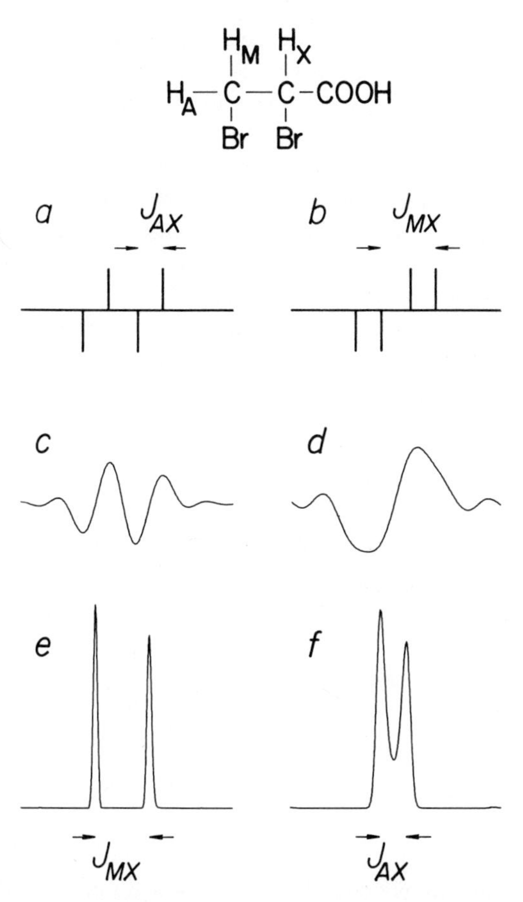

Fig. 3.17. Cross-sections through COSY spectra of 2,3-dibromopropionic acid. (a,b) Schematic spectrum showing anti-phase active splittings and in-phase passive splittings. (c,d) Conventional processing of data with severe truncation in the t_1 domain. (e,f) Maximum entropy reconstruction with J-deconvolution of the active splittings. (Courtesy of J. A. Jones and D. S. Grainger.)

used (the increment in t_1 is fixed by the required spectral width). In the present case, the coupling constants J_{AX} and J_{MX} are roughly 4.5 Hz and 12 Hz, so that for a 400 Hz spectral width, a minimum of about 90 t_1 values must be used. The difficulty associated with using such a relatively small number of t_1 values is that the data will be highly truncated in t_1, leading to the problems outlined in Section 3.5.3, if the spectrum is obtained by Fourier transformation. This is demonstrated by Fig. 3.17c,d, which shows the traces corresponding to Fig. 3.17a,b taken from the conventionally processed spectrum, with no apodization to attenuate the sinc-wiggles. The anti-phase components overlap and cause considerable loss of sensitivity, and the passive splittings cannot be measured reliably.

Maximum entropy, at least in the form discussed above, cannot be applied directly because the spectra, even without the sinc-wiggles, are not everywhere positive, as required by (3.5) (see Section 3.7). To avoid this problem, the transformation **R** may be used to ensure that the trial spectra do not contain negative peaks. More specifically, we use **R** to introduce the active, anti-phase splittings. Thus, trial spectra containing only passive, in-phase splittings are inverse Fourier transformed into the time domain and then multiplied by $\mathrm{i}\sin(\pi J t_1)$ where J is the active coupling constant. That this is the appropriate function may be seen by noting that its Fourier transform is an anti-phase doublet of delta functions, with splitting J. This technique is known as J-deconvolution (Jackson 1987; Delsuc and Levy 1988).

The resulting spectra (Fig. 3.17*e,f*) do indeed consist of an in-phase doublet with the passive splitting. These spectra are more intense than their Fourier transform counterparts because the lines are narrower and because the two components of the anti-phase doublet are now in phase and superimposed, rather than partially cancelling. For both of these cross-peaks, it would be difficult to determine the passive splitting from the conventional spectrum, but trivial to do so using the maximum entropy reconstructions.

3.7 A different approach

Finally we explore an alternative approach to maximum entropy processing of NMR data. Entropy is defined in terms of a probability distribution p_j:

$$S = -\sum_j p_j \ln p_j, \tag{3.16}$$

which is usually closely related to the quantity of interest. For example, in X-ray crystallography, p_j would be the distribution of electron density in the crystal; in astronomy it would be the flux of photons from the sky; for light scattering it might be a distribution of relaxation times. In all three cases, the quantities of interest are inherently real and positive and, once normalized, can be regarded as probability distributions. For coherent spectroscopy, the situation is less clear. In NMR, one is interested in a complex spectrum of resonances. Each resonance has amplitude and phase and, in general, both carry useful information. In the special case that all resonances have phase $\phi = 0$, the real part of the complex spectrum consists of positive, absorption-mode lines and may be regarded as a probability distribution: hence the entropy defined in (3.5) (Laue *et al.* 1985). All the examples presented in Sections 3.5 and 3.6 satisfy this rather restrictive condition.

A more general approach has recently been proposed (Daniell and Hore 1989). Two parallel treatments were presented—one classical, the other quantum mechanical—both based on the physics of time domain NMR. Here we outline the quantum mechanical approach. Imagine applying an arbitrary pulse sequence to a collection of spins in a magnetic field. Immediately after excitation ($t = 0$), the state of the spin system may be described by a density operator $\hat{\rho}(\omega)$. Specifically, the observable (xy) magnetization due to spins with frequency ω (neglecting proportionality constants) is

$$M_x(\omega) + \mathrm{i}M_y(\omega) = \mathrm{tr}\big[\hat{\rho}(\omega)(\hat{I}_x + \mathrm{i}\hat{I}_y)\big], \tag{3.17}$$

in which $\hat{I}_x$ and $\hat{I}_y$ are spin angular momentum operators and $\mathrm{tr}[\cdots]$ signifies the trace operation. At some later time t, the total transverse magnetization (neglecting relaxation) is given by

$$y_t = \int \big(M_x(\omega) + \mathrm{i}M_y(\omega)\big)\exp(\mathrm{i}\omega t)\,\mathrm{d}\omega. \tag{3.18}$$

This expression represents the trial data predicted by $\hat{\rho}(\omega)$. The total entropy of the spins is (von Neumann 1955)

$$S_I = -\sum_{\omega} \mathrm{tr}\left[\frac{\hat{\rho}(\omega)}{b}\ln\left(\frac{\hat{\rho}(\omega)}{b}\right)\right], \tag{3.19}$$

where once again, b has been introduced as a default parameter. The subscript refers to nuclei with spin quantum number I. (The classical treatment uses the distribution of nuclear spin orientations in place of the density operator.)

This is now a maximum entropy problem of the standard form. The density operator may be determined by maximizing S_I subject to the constraint that the predicted time domain signal (3.18) resembles the experimental signal, that is $\chi^2 = 2N_\mathrm{t}$. Now since the trial data depend on $\hat{\rho}(\omega)$ only through the the quantities $M_x(\omega)$ and $M_y(\omega)$, S_I should be maximized with constraints on the values of $M_x(\omega)$ and $M_y(\omega)$. Thus, we maximize

$$-\,\mathrm{tr}\left[\frac{\hat{\rho}(\omega)}{b}\ln\left(\frac{\hat{\rho}(\omega)}{b}\right)\right] - \beta_x(\omega)\,\mathrm{tr}\big[\hat{\rho}(\omega)\hat{I}_x\big] - \beta_y(\omega)\,\mathrm{tr}\big[\hat{\rho}(\omega)\hat{I}_y\big], \tag{3.20}$$

where $\beta_x(\omega)$ and $\beta_y(\omega)$ are Lagrange multipliers. This gives

$$\hat{\rho}(\omega) = b\exp\left(-\hat{1} - b\beta_x(\omega)\hat{I}_x - b\beta_y(\omega)\hat{I}_y\right) \tag{3.21}$$

where $\hat{1}$ is the unity operator. Combining this expression with (3.17) and (3.19) gives, after some straightforward algebra (Daniell and Hore 1989),

$$S_{1/2} = -\sum_{\omega}\left(z_\omega \ln\left(z_\omega + \left(1 + z_\omega^2\right)^{\frac{1}{2}}\right) - \left(1 + z_\omega^2\right)^{\frac{1}{2}}\right) \tag{3.22}$$

for spin $\frac{1}{2}$ nuclei (for example, protons). The quantity z_ω is proportional to the modulus of the complex magnetization:

$$z_\omega = \left|M_x(\omega) + \mathrm{i}M_y(\omega)\right|/b = \left(M_x(\omega)^2 + M_y(\omega)^2\right)^{\frac{1}{2}}/b. \tag{3.23}$$

To summarize, one seeks the complex function $M_x(\omega) + \mathrm{i}M_y(\omega)$ that maximizes $S_{1/2}$, subject to $\chi^2 = 2N_\mathrm{t}$. This is exactly equivalent to maximizing $S_{1/2}$ in (3.19) with respect to the matrix elements of $\hat{\rho}(\omega)$, again subject to constraints imposed by the data. The difference between (3.5) and (3.22) is, of course, that the latter will accommodate resonances of arbitrary phase, because of its dependence on the modulus of a complex magnetization rather than the amplitude of a real, positive spectrum. Equation (3.22) should therefore be more generally applicable.

An illustration of the use of this approach to the maximum entropy problem in NMR is again provided by a COSY experiment. Conventional processing of COSY data (involving double Fourier transformation) gives diagonal peaks with pure dispersion lineshapes and cross-peaks in positive or negative pure absorption mode. As shown in Fig. 3.18*b* for α,2,4-trichlorotoluene, the dispersion peaks on the diagonal have broad tails which overlap and obscure the interesting cross-peaks. Although there are various experimental methods for removing the diagonal peaks or changing them to pure absorption mode, they generally involve a loss in sensitivity.

Maximum entropy processing may be performed essentially as shown in Fig. 3.11. Using a complex trial spectrum, we anticipate that the diagonal peaks should appear with absorption lineshapes in the imaginary part of the spectrum, while the real part should contain only cross-peaks, again in absorption mode. The real part of the maximum entropy reconstruction of the COSY data (Fig. 3.18*c*) indeed shows a much reduced diagonal, and substantially clearer cross-peaks without loss (or gain) in sensitivity. (In fact, a small dispersion-mode contribution in this spectrum is inevitable, as discussed by Daniell and Hore (1989).)

3.8 Conclusions

In the preceding pages I have tried to show how the maximum entropy method may be used to process NMR data, to indicate where this might be profitable and to give a few comparisons with conventional processing.

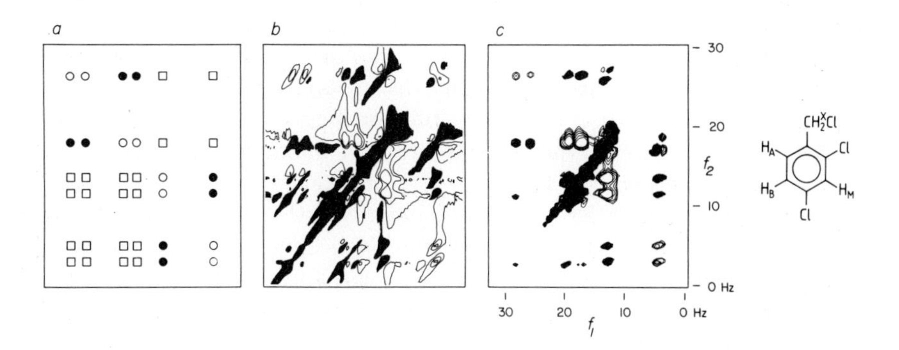

Fig. 3.18. Phase sensitive COSY spectra of α,2,4-trichlorotoluene in C_6D_6/$CDCl_3$. Only the portion of the two-dimensional spectrum due to protons A and B is shown. (*a*) Schematic spectrum. The following symbols are used—open circle: positive absorption; solid circle: negative absorption; open square: dispersion. (*b*) Experimental spectrum obtained by conventional data processing. (*c*) Maximum entropy reconstruction. In *b* and *c*, the negative contours have been filled in for clarity. (Courtesy of D. S. Grainger.)

The improvements afforded by maximum entropy in NMR are, perhaps, not as spectacular as some discussed elsewhere in this book. The reasons for this are two-fold. First, the technology of NMR hardware and the ingenuity of NMR spectroscopists are sufficiently advanced that it is usually possible to obtain data that are not grossly distorted and that are reasonably well suited to conventional processing methods. Second, dramatic improvements to the quality of a spectrum by data processing, of whatever kind, are usually only possible when noise is essentially absent, a state of affairs seldom encountered in NMR.

Despite some degree of success in avoiding the pitfalls associated with Fourier transformation of distorted and/or incomplete signals, the maximum entropy method does not have all the desirable features of the Fourier transform. First, it is typically about a hundred times slower than conventional processing and usually requires between 5 and 10 times more computer memory. These limitations will become less restrictive as computer technology advances and, anyway, if one has spent 20 hours acquiring data (a typical time for a two-dimensional NMR experiment), it does not seem unreasonable to devote a few more hours to data processing in order to get the most out of the data. Second, maximum entropy is not as convenient as Fourier transformation. It is not yet routinely available in the software supplied by NMR instrument manufacturers, and although it is not parametric (one requires no prior knowledge, or prejudice, about the number of resonances, their positions etc.) some thought is required in choosing the default parameter b, measuring the noise level σ and calculating a weighting function if resolution enhancement is required. Third, the method is non-linear and so one must expect some distortion in the relative intensities of resonances, especially if they have different line widths. Also, one needs to exercise great care in the interpretation of maximum entropy spectra because of the absence, or near absence, of baseline noise, which can no longer be used as a rule of thumb to judge whether a small splitting or intensity difference is likely to be genuine. Non-linear data processing has been likened (Jaynes 1983) to providing a carpenter with a sophisticated power tool when previously he had only hand tools. He may use it to make more beautiful furniture or he may cut off his thumb. It depends on the carpenter.

In summary, maximum entropy data processing is no substitute for careful, well-designed experiments. It is not a panacea for poor sensitivity. It *does* have something to offer, though, when data are incomplete or distorted in some well-defined way.

Acknowledgements

I am indebted to Ray Freeman and Geoff Daniell who taught me about maximum entropy, to Geoff Daniell who kindly provided the maximum entropy algorithms used here and to David Grainger, Jonathan Jones and William Broadhurst for many illuminating discussions and, not least, for Figs 3.6, 3.17 and 3.18.

References

Aue, W. P., Bartholdi, E., and Ernst, R. R. (1976). Two dimensional spectroscopy. Application to nuclear magnetic resonance. *Journal of Chemical Physics*, **64**, 2229–46.

Bartholdi, E. and Ernst, R. R. (1973). Fourier spectroscopy and the causality principle. *Journal of Magnetic Resonance*, **11**, 9–19.

Bloch, F. (1956). Dynamical theory of nuclear induction II. *Physical Review*, **102**, 104–35.

Bracewell, R. N. (1965). *The Fourier transform and its applications.* McGraw-Hill, New York.

Brigham, E. O. (1974). *The fast Fourier transform.* Prentice-Hall, Englewood Cliffs, New Jersey.

Cox, S. J. and Styles, P. (1980). Towards biochemical imaging. *Journal of Magnetic Resonance*, **40**, 209–12.

Daniell, G. J. and Hore, P. J. (1989). Maximum entropy and NMR—a new approach. *Journal of Magnetic Resonance*, **84**, 515–36.

Davies, S. J., Bauer, C., Hore, P. J., and Freeman, R. (1988). Resolution enhancement by nonlinear data processing. 'Hogwash' and the maximum entropy method. *Journal of Magnetic Resonance*, **76**, 476–93.

Delsuc, M. A. and Levy, G. C. (1988). The application of maximum entropy processing to the deconvolution of coupling patterns in NMR. *Journal of Magnetic Resonance*, **76**, 306–15.

Derome, A. E. (1987). *Modern NMR techniques for chemistry research.* Pergamon, Oxford.

Ernst, R. R. (1966). Sensitivity enhancement in magnetic resonance. *Advances in Magnetic Resonance*, **2**, 1–135.

Ernst, R. R., Bodenhausen, G., and Wokaun, A. (1987). *Principles of nuclear magnetic resonance in one and two dimensions.* Oxford University Press.

Freeman, R. (1987). *A handbook of nuclear magnetic resonance.* Longman, Harlow.

Harris, R. K. (1983). *Nuclear magnetic resonance spectroscopy.* Pitman, London.

Hoch, J. C. (1985). Maximum entropy signal processing of two-dimensional NMR data. *Journal of Magnetic Resonance*, **64**, 436–40.

Hoch, J. C., Stern, A. S., Donoho, D. L., and Johnstone, I. M. (1990). Maximum entropy reconstructions of complex (phase sensitive) spectra. *Journal of Magnetic Resonance*, **86**, 236–46.

Hore, P. J. (1985). NMR data processing using the maximum entropy method. *Journal of Magnetic Resonance*, **62**, 561–7.

Hore, P. J. and Daniell, G. J. (1986). Maximum entropy reconstruction of rotating-frame zeugmatography data. *Journal of Magnetic Resonance*, **69**, 386–90.

Hoult, D. I. (1979). Rotating frame zeugmatography. *Journal of Magnetic Resonance*, **33**, 183–97.

Jackson, R. A. (1987). Application of the maximum entropy method to the analysis of electron spin resonance spectra. *Journal of Magnetic Resonance*, **75**, 174–8.

Jaynes, E. T. (1983). *Papers on probability, statistics and statistical physics*, Synthese Library, Vol. 158 (ed. R. D. Rosenkrantz), pp. 250–1. Reidel, Dordrecht.

Jones, J. A. and Hore, P. J. (1990). The maximum entropy method and Fourier transformation compared. *Journal of Magnetic Resonance*, in press.

Laue, E. D., Skilling, J., Staunton, J., Sibisi, S., and Brereton, R. G. (1985). Maximum entropy method in nuclear magnetic resonance spectroscopy. *Journal of Magnetic Resonance*, **62**, 437–52.

Laue, E. D., Mayger, M. R., Skilling, J., and Staunton, J. (1986). Reconstruction of phase-sensitive two-dimensional NMR spectra by maximum entropy. *Journal of Magnetic Resonance*, **68**, 14–29.

Lindon, J. C. and Ferrige, A. G. (1980). Digitization and data processing in Fourier transform NMR. *Progress in Nuclear Magnetic Resonance Spectroscopy*, **14**, 27–66.

Mazzeo, A. R. and Levy, G. C. (1986). Parameterization of maximum entropy calculations for NMR spectroscopy. *Computer Enhanced Spectroscopy*, **3**, 165–70.

Morris, P. G. (1986). *Nuclear magnetic resonance imaging in medicine and biology.* Oxford University Press.

Neumann, J. von (1955). *Mathematical foundations of quantum mechanics.* Princeton University Press, New Jersey.

Ni, F., Levy, G. C., and Scheraga, H. (1986). Simultaneous resolution enhancement and noise suppression in NMR signal processing by combined use of maximum entropy and Fourier self-deconvolution methods. *Journal of Magnetic Resonance*, **66**, 385–90.

Redfearn, J. (1984). *The Times.* 17 October, London.

Sibisi, S. (1983). Two-dimensional reconstructions from one-dimensional data by maximum entropy. *Nature*, **301**, 134–6.

Sibisi, S., Skilling, J., Brereton, R. G., Laue, E. D., and Staunton, J. (1984). Maximum entropy signal processing in practical NMR spectroscopy. *Nature*, **311**, 446–7.

Skilling, J. and Bryan, R. K. (1984). Maximum entropy image reconstruction: general algorithm. *Monthly Notices of the Royal Astronomical Society*, **211**, 111–24.

Stephenson, D. S. (1988). Linear prediction and maximum entropy methods in NMR spectroscopy. *Progress in Nuclear Magnetic Resonance Spectroscopy*, **20**, 515–626.

Styles, P., Scott, C. A., and Radda, G. K. (1985). A method for localizing high-resolution NMR spectra from human subjects. *Magnetic Resonance in Medicine*, **2**, 402–9.

Wüthrich, K. (1986). *NMR of proteins and nucleic acids.* Wiley, New York.

4

Enhanced information recovery in spectroscopy using the maximum entropy method

S. Davies, K. J. Packer, A. Baruya and A. I. Grant

Abstract

The maximum entropy method for the reliable extraction of information from spectroscopic data is outlined. Examples are shown from Raman and nuclear magnetic resonance spectroscopies, with particular reference to the selection and parameterization of suitable lineshape functions for deconvolution. An account is also given of our experiences with a proposed definition of the entropy function different from the usual form.

4.1 Introduction

The high cost of modern analytical instrumentation and the availability of powerful, low cost computers has stimulated a great deal of academic and commercial interest in the general area of off-line processing of spectroscopic data. A plethora of mathematical techniques is available (see, for example, the review by Stephenson (1988)) which provide information in a variety of forms ranging from conventional frequency spectra to the parameters of models used to interpret the raw data. Advanced data processing techniques have developed to become useful tools in the armoury of the spectroscopist—many spectroscopic studies rely heavily upon some form of modelling or deconvolution.

Our activities in this area have centred around the development of practical applications of the maximum entropy method (MaxEnt[1]) in Raman and nuclear magnetic resonance (NMR) spectroscopies.

In order to be viable from a commercial point of view, any data processing method must satisfy some basic criteria:

[1] *Editors' Note.* Or MEM which is the acronym preferred at BP.

1. What new information is available?
2. Is the extra information useful?
3. Is the technique quantitative?
4. Does the result justify the additional computational cost?

This chapter describes our experiences in implementing a MaxEnt package (based around the Maximum Entropy Data Consultants Ltd (MEDC) `MEMSYS1` core routines) for the routine deconvolution of spectroscopic data, addressing the criteria listed above.

4.2 Background

Spectroscopic data are necessarily incomplete due to the limiting constraints of digital data recording and finite acquisition time. The traditional approach to spectroscopic data processing is to ignore any known imperfections in the data, set to zero any unmeasured data points, and then proceed as if the data were perfect. Thus, the parameters of interest (frequencies, intensities, phases and line-widths) may be distorted.

A simple example of the potential problems with conventional data processing occurs in Fourier spectroscopy (for example, NMR spectroscopy). Consider an analogue time domain signal $f(t)$ which is an exponentially damped sinusoid with intensity I, phase ϕ, decay constant T_2, and frequency ν_0:

$$f(t) = I \exp(2\pi \mathrm{i} \nu_0 t) \exp(-t/T_2) \exp(\mathrm{i}\phi) \qquad (t \geq 0). \tag{4.1}$$

Analytical (continuous) Fourier transformation is defined by the expression

$$F(\nu) = \int_{-\infty}^{\infty} f(t) \exp(-2\pi \mathrm{i} \nu t) \, \mathrm{d}t. \tag{4.2}$$

Thus, the ideal frequency domain spectrum $F(\nu)$ is given by

$$F(\nu) = I T_2 \exp(\mathrm{i}\phi) \big(1 + 2\pi \mathrm{i}(\nu - \nu_0) T_2\big)^{-1}. \tag{4.3}$$

The real and imaginary parts of (4.3) correspond to Lorentzian absorption and dispersion signals, respectively, when $\phi = 0$.

In practice, however, the analogue signal is sampled at a finite rate, yielding a discrete time series which is processed using a discrete Fourier transform, defined by

$$F\left(\frac{m}{N\Delta t}\right) = \sum_{k=0}^{N-1} f(k\Delta t) \exp(-2\pi \mathrm{i} m k/N), \tag{4.4}$$

where Δt is the sampling interval and N the number of sampled points. The integer m takes values from 0 to $N-1$. A further departure from ideal behaviour is caused by the fact that the detection circuitry cannot usually respond immediately to the analogue signal, resulting in a delay Δx, the so-called dead-time, before acquisition commences. The discrete frequency domain spectrum thus has the form

$$F\left(\frac{m}{N\Delta t}\right) = C(1-y^N)/(1-y), \tag{4.5}$$

where

$$y = \exp\Big(\big(2\pi \mathrm{i}(m_0 - m)/N\big) - \Delta t/T_2\Big) \tag{4.6}$$

and

$$C = I\exp(\mathrm{i}\phi)\exp(2\pi \mathrm{i} m_0 \Delta x/N\Delta t)\exp(-\Delta x/T_2), \tag{4.7}$$

with $m_0 = \nu_0 N\Delta t$. Equation (4.5) is clearly not equivalent to (4.3). In the limit that $N\Delta t$ (the data acquisition time) is large compared with T_2, the sampling frequency $1/\Delta t$ large and the dead-time Δx small, then (4.5) becomes the discrete analogue of (4.3). If these conditions are not met, even the apparently simple procedure of producing a digital record of an analogue signal may introduce distortions. The presence of the dead-time Δx in (4.5) gives rise to a frequency-dependent phase-shift. Failure to sample sufficiently rapidly may lead to the well-known phenomenon of 'aliasing', where signals which have frequencies outside the range encompassed by the sampling rate appear in the spectrum at a lower frequency.

Spectroscopic data are affected, in general, by a whole variety of imperfections:

1. finite bandwidth caused by the finite sampling rate,
2. truncation of the data caused by the finite data acquisition time,
3. distortion introduced by the detection system—the experimental data are convolved with the transfer function of the spectrometer,
4. all experimental data are corrupted to some degree by noise which consists of both thermal noise, usually caused by the random motion of electrons in the detection circuitry, and digital noise caused by the finite word length used to represent the analogue signal.

Any technique which manipulates the experimental data directly will incorporate these imperfections into the resultant spectrum. Perhaps the most common example is shown in Fig. 4.1, where a data set consisting of a truncated, exponentially damped sinusoid is used to represent the 'experimental' data (Fig. 4.1*a*). Fig. 4.1*b* shows the discrete Fourier transform of this signal. Fig. 4.1*c* shows the discrete Fourier transform after zero-filling three times (zero-filling once means adding an equal number of zeros to the

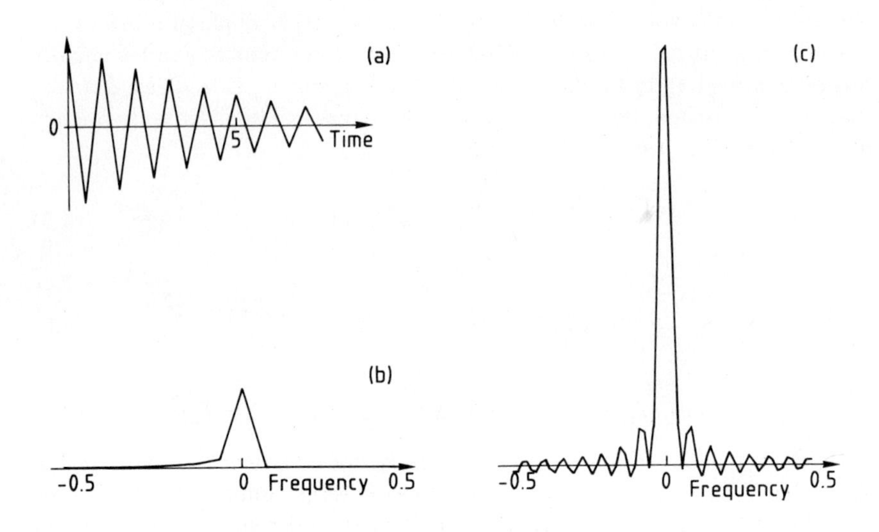

Fig. 4.1. Spectra obtained by discrete Fourier transformation of truncated time domain data (*a*). The frequency spectrum with no zero-filling is shown in *b* and after zero-filling three times in *c*.

end of the data). The practice of zero-filling is a very useful way of interpolating extra points into the spectrum, without acquiring extra data. However, it leads to characteristic 'sinc-wiggles' when the data are truncated, rendering interpretation difficult in the presence of more than one signal, particularly when the dynamic range is large. The sinc-wiggles are not readily apparent in Fig. 4.1*b* because the data points coincide approximately with the zero-crossings of the sinc function ($\mathrm{sinc}(x) = \sin(\pi x)/(\pi x)$).

Most advanced data processing methods do not perform any manipulation upon the raw data, but calculate theoretical data, incorporating any appropriate prior knowledge of the experimental conditions. The theoretical data are then fitted to the experimental data via some appropriate constraint, yielding either model parameters or an idealized spectrum.

4.3 The maximum entropy method (MaxEnt)

The basic problem facing the spectroscopist may be cast in the general form

$$D_i = \sum_{k=1}^{N} O_{ik} f_k + \sigma_i, \tag{4.8}$$

where the D_i are the (imperfect and incomplete) experimental data, the f_k represent the ideal (perfect) spectrum of the sample (or more generally, the 'image'), the matrix $\mathbf{O}$ represents the instrumental response or blurring function and the σ_i represent the noise on each digitized point. It should be noted that this simple formulation is strictly appropriate only for unidimensional problems (although the extension to higher dimension is essentially trivial). It is also assumed that the observed noise is independent of the image $\boldsymbol{f}$.

Thus, the aim is to obtain the best possible estimate of $\boldsymbol{f}$ using all available information about the physics of the experiment and any likely imperfections. This information is used to build a model of the blurring function $\mathbf{O}$.

The image $\boldsymbol{f}$ cannot be recovered by direct inversion of (4.8), even if the matrix $\mathbf{O}$ is known, owing to the presence of the noise and the finite nature of $\boldsymbol{D}$. However, the maximum entropy method offers a very powerful way of calculating a defensible reconstructed image $\boldsymbol{f}$. The theoretical aspects of the technique are discussed in detail in Chapters 1 and 2, so only a very brief description of the method will be given here.

The problem is that the incomplete noisy data $\boldsymbol{D}$ may be equally consistent with an infinite variety of different images $\boldsymbol{f}$. In the absence of any independent information about $\boldsymbol{f}$, it is difficult (or impossible) to defend the arbitrary selection of any one of the possible images as the 'best' estimate of $\boldsymbol{f}$. The principle of maximum entropy provides an escape route from this dilemma, by specifying that, of those images consistent with the measured data, the maximally non-committal choice is to select the image with the minimum information content, or, equivalently, the maximum entropy. This choice is least likely to lead to over-interpretation of the available data.

Thus, the maximum entropy method selects that image with maximum configurational entropy, defined by

$$S = -\sum_{i=1}^{N} p_i \ln \frac{p_i}{w_i}, \tag{4.9}$$

where p_i is the normalized, weighted spectral intensity at frequency i, defined by

$$p_i = \frac{w_i f_i}{\sum_{j=1}^{N} w_j f_j}, \tag{4.10}$$

and where w_i is the weight assigned to frequency i. For most spectroscopic problems, we have no prior prejudice with regard to the expected intensities in frequency space, and so the w_i can be assigned to be unity for all i.

The definition of the entropy has been the subject of much discussion, but the only definition for which there is any *a priori* theoretical justification is that containing a term of the form $p \ln p$ (the derivation of a suitable form for the entropy is discussed in Chapter 2). It should be noted that, with the above definition of entropy, MaxEnt is only applicable in situations where the image consists of positive, additive signals, because the term within the logarithm of (4.9) must be positive. Thus, this simple approach will fail in cases where the image consists of complex signals of varying phase.

The procedure adopted to calculate the maximum entropy spectrum is to start with a 'trial' spectrum which is usually flat and featureless to avoid bias (although if reliable prior information is available, known spectral features may be included at this stage). The trial spectrum is transformed to give mock experimental data, using a form of **O** appropriate for the particular spectroscopy. The mock data are then compared with the experimental data via an appropriate consistency test. For most spectroscopic problems the noise statistics are approximately Gaussian with constant variance σ^2, so a normalized chi-squared statistic is appropriate, defined by

$$C(f) = \frac{1}{N}\sum_i \frac{\left(D_i - F_i\right)^2}{\sigma_i^2}, \tag{4.11}$$

where the F_i are the mock experimental data and the σ_i are taken to be equal to σ for all values of i. The chi-squared surface $C(f) = 1$ in image space is convex. The entropy surface is similarly shaped, such that there is only one maximum entropy solution consistent with both the data and the specified model for **O**. The solution lies at the extremum of

$$Q(\lambda, \boldsymbol{f}) = S(\boldsymbol{f}) - \lambda C(\boldsymbol{f}), \tag{4.12}$$

where λ is a Lagrange multiplier. The maximum entropy solution is approached iteratively by creating a subspace spanned by three search directions:

1. the gradient of the entropy,
2. the gradient of the consistency test in image space,
3. a combination of the two in the form of a gradient of Q.

Suitable combinations of these search directions are used to calculate increments for the image. On each iteration, the whole image is incremented. Fig. 4.2 shows a schematic diagram of the iteration process. More detail of the operational algorithm can be found in papers by Skilling and Gull (Skilling 1984; Skilling and Gull 1985).

The `MEMSYS` package, supplied by MEDC, requires the user to provide two routines (called `OPUS` and `TROPUS`) to interconvert between the image and data spaces. The package may, therefore, be applied in a wide variety

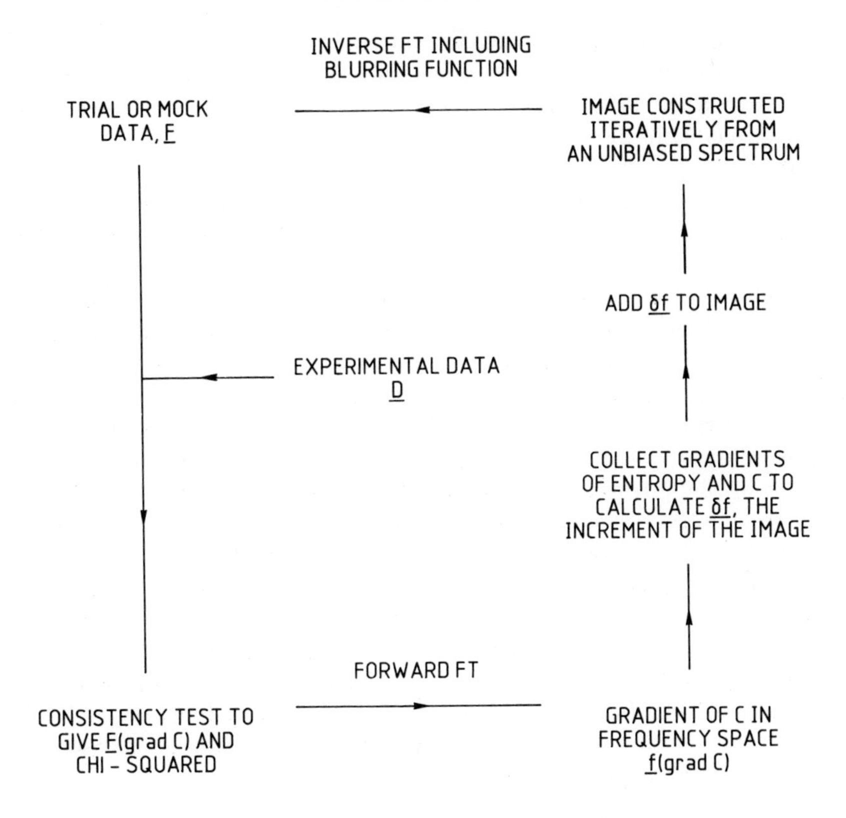

Fig. 4.2. Flow diagram of a MaxEnt iteration process.

of applications, with the choice of blurring function **O** left entirely to the user.

4.4 Applications

The physics behind most routine forms of spectroscopy is well understood, so that a great deal of information about theoretical lineshapes and likely distortions of the data is available. The relationship between the image and data spaces is typically fairly straightforward (for example, a Fourier transform) and the blurring function usually takes the form of a simple convolution (often frequency independent). Thus, the computation involved in the processing of unidimensional data is not too time consuming. For a typical experimental data set consisting of a few thousand points, a realistic MaxEnt reconstruction is usually obtained in ten minutes to half an hour of CPU time on a VAXstation 3100.

A typical application of MaxEnt in spectroscopic data processing is to obtain resolution enhancement without the simultaneous degradation of signal-to-noise ratio which occurs with linear deconvolution procedures. Resolution enhancement is achieved by accurate choice of the blurring function $\mathbf{O}$.

The effect of the spectrometer imperfections is to smear out the true signals, yielding spectra with broadened lines. Application of the algorithm with the identity matrix as blurring function yields a MaxEnt reconstruction with broad lines since the effect of instumental and other broadening has been omitted. Convolution of the trial data with a line broadening function before the chi-squared test forces a maximum entropy solution with lines narrowed relative to the reconstruction using the identity matrix. A suitable starting point for the choice of $\mathbf{O}$ is to take note of the fact that the widths of many spectroscopic lines are dominated by instrumental factors and not by the intrinsic line widths. Thus, a good choice for the blurring function will often be some lineshape function with a line width matched approximately to the observed line widths in the experimental data. 'Fine tuning' of the lineshape function can yield reconstructions with widths approaching (or sometimes less than) the natural line widths. The validity of a result where the line width is less than that allowed by the physics of the situation is, of course, somewhat dubious, as the deconvolution is being over-driven. In this case a less severe broadening function is clearly required. Thus, the optimum choice of broadening function depends upon the intuition and experience of the spectroscopist.

4.4.1 Raman spectroscopy

When light is scattered by a molecule, its frequency may be unaltered (elastic scattering) or shifted (inelastic scattering). In the latter case, any frequency shift (which may be either positive or negative) depends upon the fundamental vibrational frequencies of the scattering species. Raman spectroscopy is the study of such inelastic scattering, and may be applied to the study of gases, liquids and, most importantly, to surfaces.

The basic Raman spectroscopy experiment consists of illuminating a sample with a laser and monitoring the frequency and intensity of the scattered radiation. Laser Raman spectra usually possess only modest signal-to-noise ratio, owing to the weakness of the spontaneous Raman effect: most of the light is elastically scattered. Of every billion photons that strike the sample, only about 1 to 500 scattered photons will show frequency shifts due to the Raman effect. High throughput monochromators are used to compensate for the inherently low sensitivity of the technique, but wide slit widths degrade the resolution. This effect is compounded by the large natural line-widths which occur when the sample is of a condensed phase. Thus, in a typical Raman experiment, a compromise has

Table 4.1. Symmetric breathing mode frequencies of CCl_4.

Band	Frequency (cm^{-1})	MaxEnt intensity	Theoretical intensity
$C\,^{37}Cl_4$	Not observed	N/A	0.4%
$C\,^{37}Cl_3\,^{35}Cl_1$	452.0	6%	4.7%
$C\,^{37}Cl_2\,^{35}Cl_2$	456.4	20%	21.1%
$C\,^{37}Cl_1\,^{35}Cl_3$	459.4	41%	42.2%
$C\,^{35}Cl_4$	462.4	31%	31.6%

to be struck between adequate signal-to-noise ratio and spectral resolution, which makes the reconstruction of Raman spectra an ideal task for MaxEnt. A further complication is that those spectra recorded at visible wavelengths occasionally exhibit very high background photon counts, owing to fluorescence excited by the laser. The Raman spectrum is then seen to sit on a broad non-uniform 'hump'. In the worst cases, the fluorescence may totally obscure the spectrum.

The $459\,cm^{-1}$ band in the Raman spectrum of carbon tetrachloride (CCl_4) is often used as a resolution test for Raman spectrometers. This line corresponds to the totally symmetric breathing mode of the molecule: the four carbon–chlorine bonds vibrate with the same frequency and phase. Naturally occurring chlorine is a mixture of the isotopes ^{35}Cl and ^{37}Cl in the ratio of approximately 3 : 1. Thus, the breathing mode vibration is split into five discrete components, which correspond to the five possible isotopic combinations within the molecule. Fig. 4.3 shows the $459\,cm^{-1}$ band of CCl_4 recorded using a slit width of $5\,cm^{-1}$. The resolution is rather poor; the band could be accurately described as an asymmetric lump. Fig. 4.4 shows the MaxEnt reconstruction of the spectrum using a Lorentzian lineshape, with a full width at half maximum (FWHM) of $4.5\,cm^{-1}$, as the blurring function $\mathbf{O}$. Fig. 4.4 is the best MaxEnt reconstruction we could obtain from the data of Fig. 4.3 (Grant and Packer 1989). Some fine structure has been resolved, but there is still considerable overlap within the spectrum. Fig. 4.5 shows the same band, but recorded with a much narrower slit width of $0.9\,cm^{-1}$: the signal-to-noise ratio is lower than in Fig. 4.3, but the resolution is considerably better, with three components of the band visible. Fig. 4.6 shows the MaxEnt reconstruction from the data of Fig. 4.5 using a Lorenztian blurring function with FWHM of $0.8\,cm^{-1}$. The enhancement in resolution is striking. Four components are clearly visible; the fifth component is masked by a broad hump at the low frequency end of the spectrum. The intensities derived from the MaxEnt spectrum and the corresponding theoretical intensities are shown in Table 4.1. MaxEnt's numerical estimates of relative spectral intensities

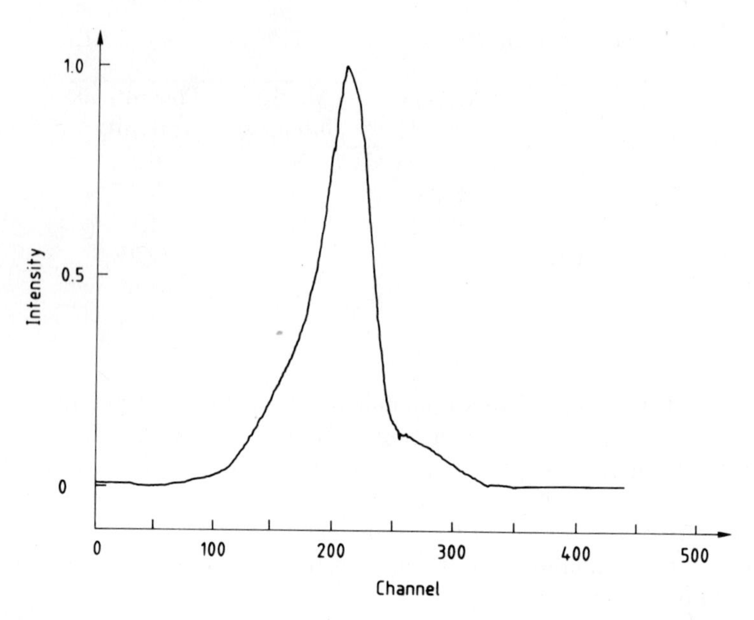

Fig. 4.3. 459 cm^{-1} band of CCl_4 measured using a slit width of 5 cm^{-1}.

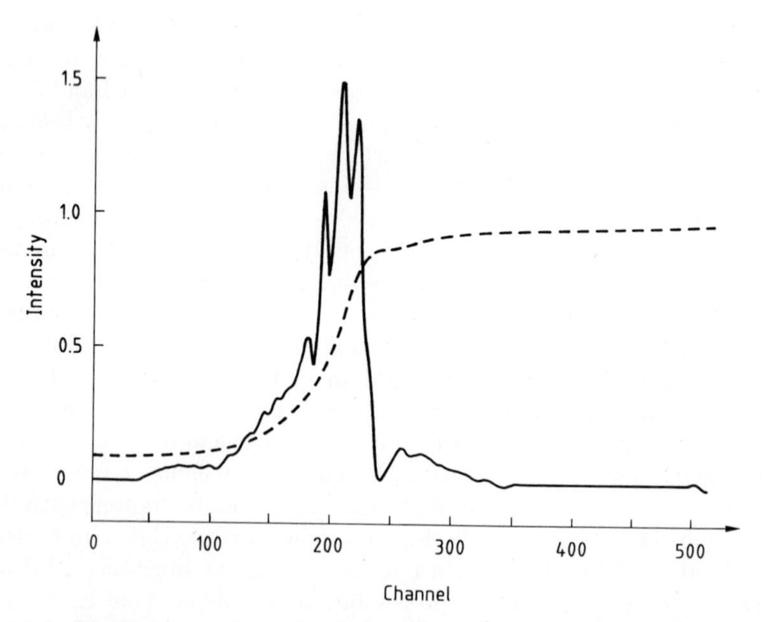

Fig. 4.4. MaxEnt reconstruction of the 459 cm^{-1} band of CCl_4 using the data of Fig. 4.3. The blurring function is a Lorentzian with FWHM of 4.5 cm^{-1}. Also shown is the integrated spectrum (dashed line).

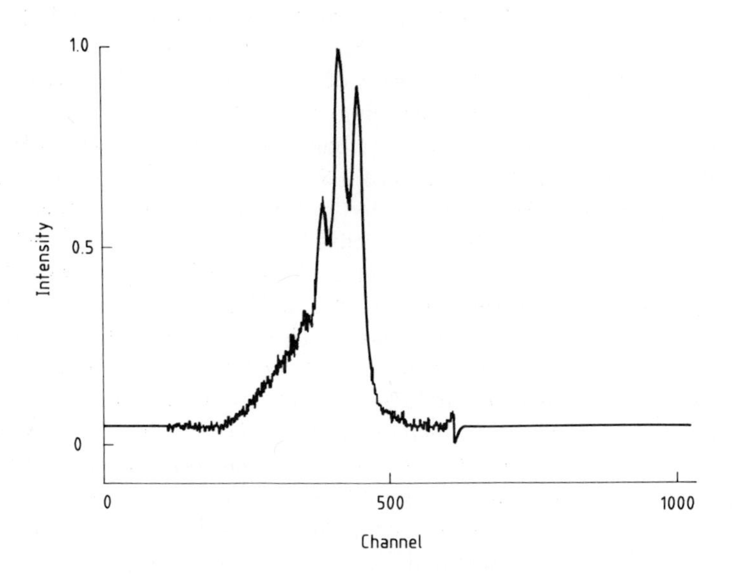

Fig. 4.5. 459 cm^{-1} band of CCl_4 measured using a slit width of 0.9 cm^{-1}.

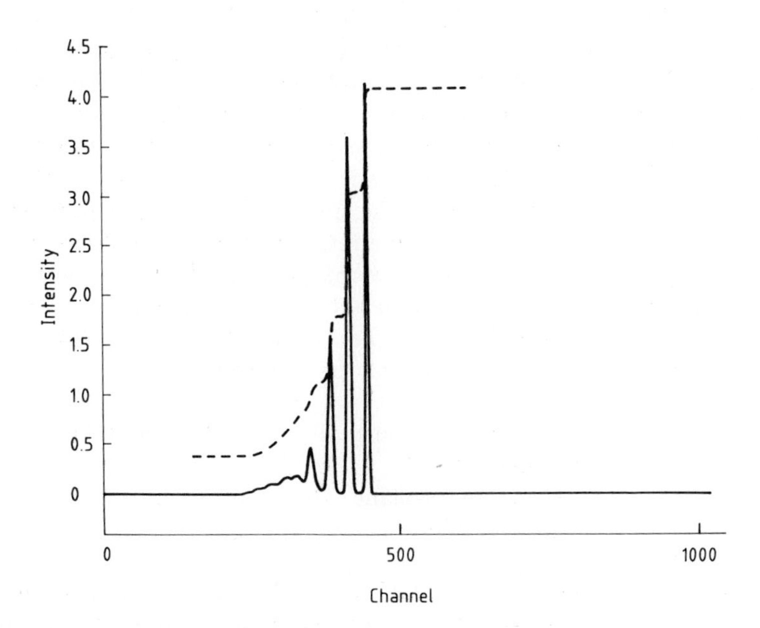

Fig. 4.6. MaxEnt reconstruction of the 459 cm^{-1} band of CCl_4 using the data of Fig. 4.5. The blurring function is a Lorentzian with FWHM of 0.8 cm^{-1}. Also shown is the integrated spectrum (dashed line).

appear to be reliable. Of course, there is a problem here in that there are no error bars[2] for the intensities within the reconstructed spectrum. The only way available for the assessment of errors is to examine the residuals of the fitting process for any remaining structure.

Probably the most significant (albeit intuitively obvious) feature to emerge from the comparison of Figs 4.4 and 4.6 is that the degree of resolution enhancement achievable is limited by the quality of the original data, so that MaxEnt will never be a substitute for good experimental technique. In fact, the MaxEnt reconstruction merely shows what defensible information is contained within the experimental data, given the prior knowledge built into the blurring function. Thus, with inadequate data, genuine features may not appear in the reconstruction—that of Fig. 4.6 illustrates the point, as follows. There is, in fact, a broad 'hot band' (an overtone of a lower frequency resonance) which overlaps the symmetric stretching mode of CCl_4 (see, for example, Loader (1970)). The reconstruction does not accurately reveal this response because there is insufficient evidence for the signal in the raw data. Such behaviour has been the source of much false criticism of MaxEnt (aided and abetted by some of its proponents) because the MaxEnt reconstruction should not be thought of as the 'true' frequency spectrum of the sample, but as an intensity versus frequency representation of the information for which there is evidence in the data. Non-appearance of a 'known' signal simply shows that the raw data are not good enough.

A spectacular example of the deconvolution power of MaxEnt arises in a study of graphite/$FeCl_3$ intercalates. Graphite is well known for its layered structure of carbon sheets and its high electrical conductivity. Intercalation of small molecules between the layers is used to modify the electrical and chemical properties. The object of the spectroscopic study is to determine the stage of intercalation (that is, the ratio of free graphite layers to intercalated layers). Fig. 4.7 shows the raw data and the MaxEnt reconstruction using a Lorentzian blurring function with FWHM of $25\,cm^{-1}$. This result is guaranteed to generate disbelief wherever it appears! Nevertheless, the two features in the reconstruction may be assigned:

$1582\,cm^{-1}$	Graphite in-plane vibration from inter-layer graphite sheets,
$1617\,cm^{-1}$	Graphite in-plane vibration from bounding layer graphite sheets.

The natural graphite peak is at $1580\,cm^{-1}$ so the shifted peak is presumably due to the effects of intercalation. From the reconstruction the material is assigned to be a stage 3 intercalate. Such materials oxidize very rapidly in

[2] *Editors' note.* For a reinterpretation of the method which allows errors to be estimated, see Chapter 2.

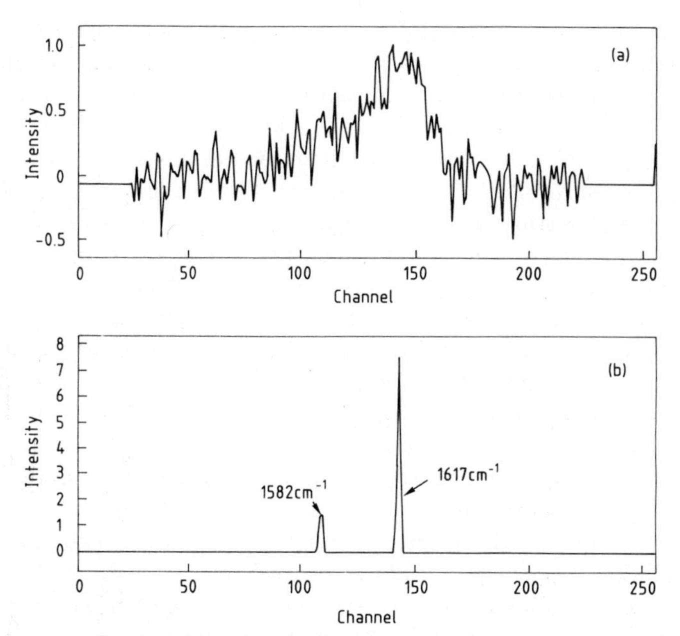

Fig. 4.7. (*a*) Raman spectrum of graphite/$FeCl_3$ intercalate. (*b*) MaxEnt reconstruction with a Lorentzian blurring function with FWHM of $25\,cm^{-1}$.

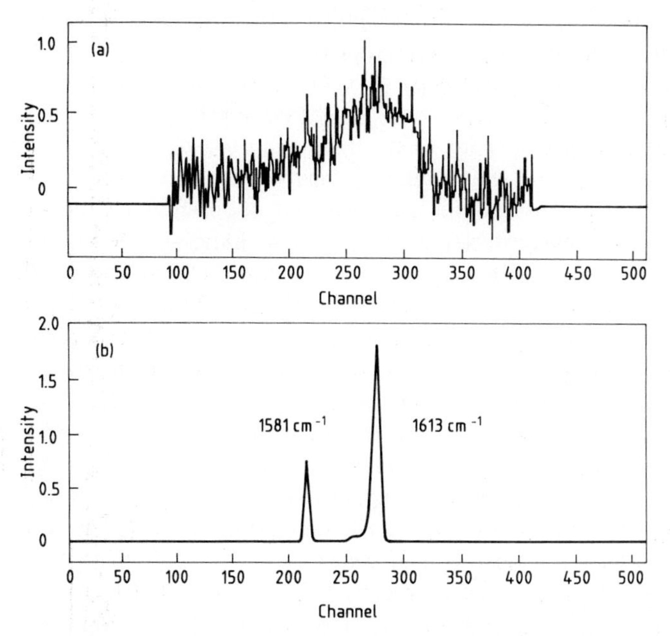

Fig. 4.8. (*a*) Raman spectrum of the graphite/$FeCl_3$ intercalate, taken five hours after the data of Fig. 4.7. (*b*) MaxEnt reconstruction of *a*.

air (some are pyrophoric) and deteriorate rapidly when exposed to oxygen at very low levels, such as those which enter the 'airtight' sample containers. A spectrum from the same sample recorded five hours later and the MaxEnt reconstruction are shown in Fig. 4.8. The results are equally dramatic. The 1581 cm^{-1} graphite peak is now increased in intensity and the feature due to intercalation is broader and shifted to 1613 cm^{-1}. This behaviour is indicative of a stage 3/4 or a stage 4 intercalate, illustrating that the sample has undergone considerable oxidation. Such deductions would not be possible from the raw data. However, the question of error bars for the image remains unanswered. Quantitative interpretation of these results would, therefore, ideally require some independent means of verification.

The reliable performance of the MaxEnt scheme is, at least to some extent, dependent upon the quality of the spectroscopist. That is to say, the prior knowledge and any assumptions imposed upon the data must be valid. This statement is not as restrictive as it might first seem, because the algorithm is surprisingly tolerant of incorrect estimations of the experimental response. However, in order to recover the maximum amount of information from a data record, all available prior knowledge should be built into the system. The most obvious source of inaccuracy is that the blurring function is unlikely to be accurately described by a simple functional form such as a Lorentzian or Gaussian. The obvious solution to this is to use a lineshape determined under the prevailing experimental conditions. A simple, albeit slightly artificial, demonstration of this is provided by the Raman spectrum of the sulphate ion SO_4^{2-}. The sulphate ion exhibits a strong Raman-active vibration around 980 cm^{-1}. Fig. 4.9 (dashed lines) shows a composite spectrum obtained by adding the normal sulphate Raman spectrum to itself, but shifted by 0.4 of the FWHM and attenuated to 0.75 of the original intensity. Fig. 4.9*a* (solid line) shows the best MaxEnt reconstruction achieved with a Lorentzian blurring function; a shoulder is clearly apparent on the right-hand side of the major peak. Fig. 4.9*b* (solid line) shows the reconstruction achieved using the original (slightly asymmetric) sulphate lineshape as the blurring function; the two lines are now clearly resolved. The experimental sulphate lineshape clearly contains extra information about the imperfections in the data acquisition process and therefore yields an improved reconstruction. The important point here is, of course, that the blurring function is noisy, and so this sort of deconvolution procedure is best confined to spectra where the signal-to-noise ratio is high. It is probably best not to examine this artificial case in too much detail as the effects of correlations between the points in the 'raw' data have not been considered.

A more realistic (and useful) example is provided by a spectroscopic study of acrylate copolymerization reactions. Raman spectroscopy can readily be applied to the study of such reactions. The problem is that the reaction kinetics need to be obtained from the rate of disappearance of

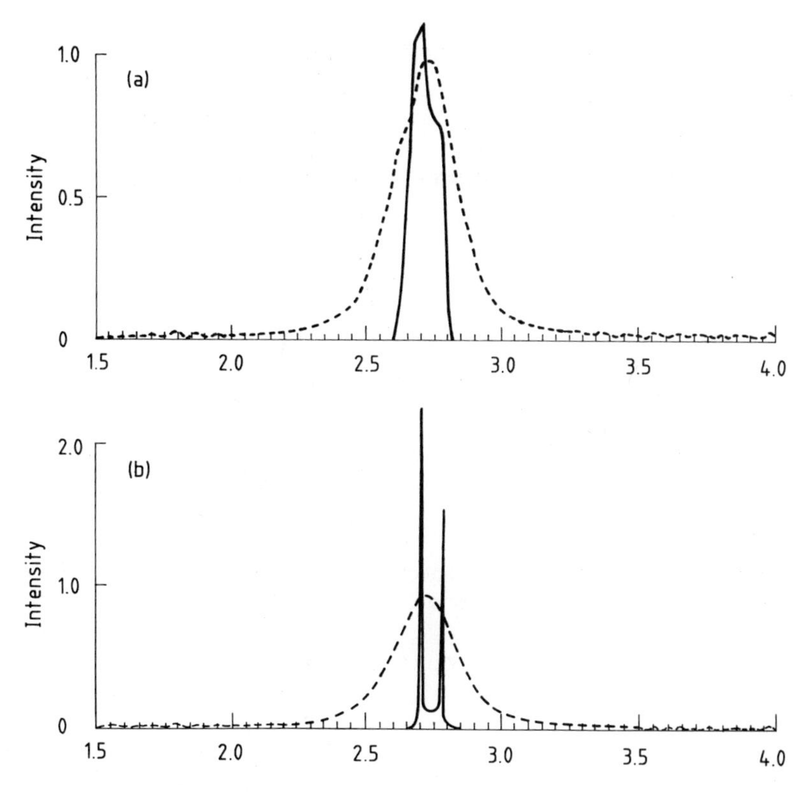

Fig. 4.9. (*a*) Composite spectrum (dashed line) constructed from the Raman spectrum of the sulphate ion, as described in the text, and the MaxEnt reconstruction (solid line) using a Lorentzian blurring function. (*b*) The same spectrum (dashed line) and the MaxEnt reconstruction (solid line) using the original sulphate spectrum as the blurring function.

the C=C stretching mode, since it is through the saturation of this grouping that polymerization takes place. Unfortunately, the stretching vibrations for acrylates and methacrylates tend to be very close in frequency, so it is difficult to quantify the changes in one such compound in the presence of another. This is an ideal application for MaxEnt. A further complication is that the vinyl vibration at 1640 cm^{-1} has a non-Lorentzian, asymmetric profile. Thus, an experimentally determined profile recorded from one of the neat monomers has been employed as the blurring function. Fig. 4.10 shows a typical raw data set, taken from a 3 : 1 ethylhexyl acrylate/ n-butyl acrylate monomer mix. The neat monomer n-butyl acrylate spectrum is shown superimposed. The overlap is very pronounced. The n-butyl acrylate spectrum is used as the pattern match profile because it is slightly

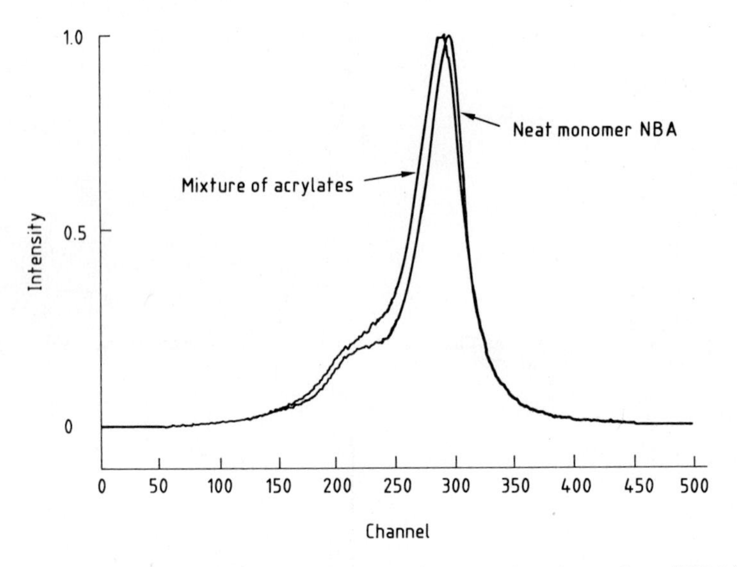

Fig. 4.10. Superimposed Raman spectra of neat n-butyl acrylate (NBA) and ethylhexyl acrylate/n-butyl acrylate monomer mix.

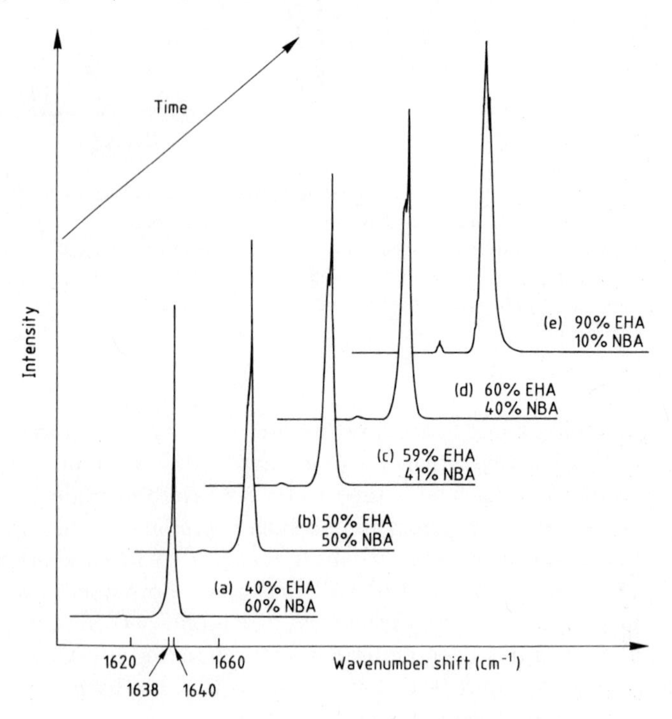

Fig. 4.11. MaxEnt reconstructions of the reacting acrylate mixture as a function of time.

narrower than the spectrum from the 2-ethyl acrylate. It is for this reason that the reconstructions shown in Fig. 4.11 display a very sharp feature, due to the n-butyl monomer, and a broader composite signal due to the 2-ethylhexyl monomer. Each reconstruction shows the actual ratios of the two monomers in the mixture. In order to quantify the results, the easiest method is probably to perform MaxEnt reconstructions for all likely ratios and to compile a look-up table of the relative intensities of the lines. The experimental technique and conditions are sufficiently stable to justify such an approach.

4.4.2 Fourier transform nuclear magnetic resonance spectroscopy (FT-NMR)

NMR spectroscopy is an extremely powerful analytical technique which may be applied to samples (solids, liquids, and gases) containing nuclei of non-zero nuclear spin. According to quantum mechanics, a nucleus of spin quantum number I has $2I + 1$ available states. Imposition of a magnetic field changes the energies of these states, and it is possible to induce transitions between them by the application of radiofrequency radiation. NMR spectroscopy is the study of how ensembles of nuclear spins interact with each other and with their environment, using radiofrequency radiation as the probe.

The wide applicability of NMR techniques stems from the sensitivity of the nuclear spins to their chemical environment. Nuclear spins in different chemical environments show small differences in their resonant frequencies, a phenomenon known as the chemical shift. A particular chemical shift is characteristic of a particular chemical environment, which makes NMR a highly versatile probe of chemical structure. In addition to the chemical shift, the nuclei may interact with each other, either directly through space (a dipole–dipole interaction) or indirectly through chemical bonds (scalar or 'J' coupling) yielding additional information about molecular shape, dynamics, and the topology of the spin coupling networks. Nuclei with spin greater than $\frac{1}{2}$ also possess a nuclear quadrupole moment which interacts with electric field gradients within the molecule.

The basic FT-NMR experiment consists of placing a sample within a strong static magnetic field and allowing it to approach equilibrium. The redistributed spins produce a net magnetization along the axis of the applied field. A short (a few microseconds) intense radiofrequency pulse of the appropriate frequency is then applied to rotate this magnetization into the plane perpendicular to the applied field. After the pulse, the magnetization is subject to a torque and precesses about the applied field, inducing an oscillating voltage in a receiver coil. The voltage is measured as a function of time (relative to a reference derived from the transmitter frequency) yielding a signal which in the simplest case is a sum of exponentially damped sinusoids: the frequencies ν_{0j} are characteristic of the various chemical environments within the sample; the line widths in the spectrum are determined by the decay constants T_{2j}^*; and the absorptive/dispersive nature of the lines by the phase angles ϕ_j:

$$S(t) = \sum_j I_j \exp(2\pi i \nu_{0j} t) \exp(-t/T_{2j}^*) \exp(i\phi_j). \qquad (4.13)$$

The time domain signal described by (4.13) is complex, with the result that the frequency spectrum is also complex. In addition the various component signals do not necessarily have the same phase: many NMR experiments produce signals with varying phases, which carry useful information as to the nature of the coupling networks. Thus, there is clearly a problem in applying the entropy definition of (4.9) to a trial frequency domain spectrum. However, the standard MaxEnt approach may be applied in NMR if a limited amount of preprocessing of the data is undertaken.

Consider the simplest NMR experiment, as described above. The first problem is that of the dead-time following the intense radiofrequency pulse; data acquisition cannot begin until the receiver circuitry has recovered. During this dead-time Δx, the magnetization precesses around the axis of the applied field giving a linear frequency-dependent phase shift $\exp(2\pi i \nu_0 \Delta x)$, where ν_0 is the resonance frequency (that is, the centre) of any line in the spectrum. Additional instrumental factors (such as the lengths of cables) give rise to a frequency-independent phase shift of the form $\exp(i\theta)$. In order to apply MaxEnt to such a spectrum the data must be manipulated so that all the lines are of the same phase. The phasing of the spectrum is achieved by multiplying the spectrum by a term of the form $\exp(2\pi i \nu y + i\theta)$, where ν is the spectral frequency and y and θ are determined by eye. This is an approximate procedure because the lines in the spectrum have a finite width; strictly each line should have only a single phase correction across the whole of its width. Thus, the wings of a line receive a different phase shift from the centre, which distorts the intensity. The effect is at its worst for broad lines where a larger phase shift is imposed across the line. In the case of overlapping lines it is clearly impossible to phase the individual lines separately without already knowing their positions, phases and intensities and also the characteristics of the noise. Additionally, the criterion of correct phasing is subjective in the presence of noise, so the operator is biasing the data prior to using MaxEnt. Despite the obvious disadvantages, this method of phasing can yield spectra where all the lines appear to be in phase (the phasing procedure is the standard method of manipulating experimental data when not using any special data processing techniques). MaxEnt reconstruction of the real (all positive) part of the spectrum can now be undertaken, as the problem caused by the entropy definition has been circumvented.

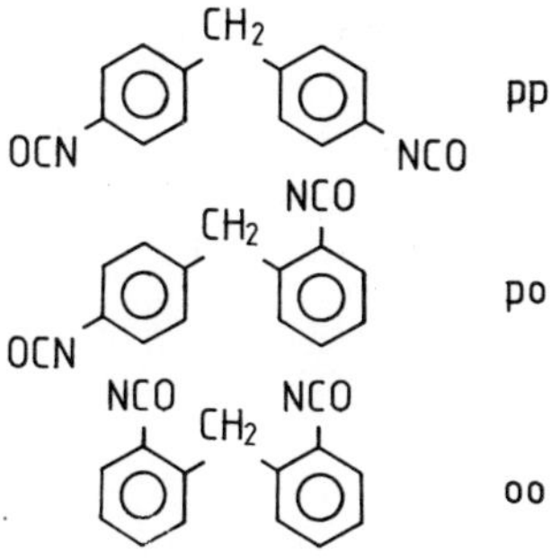

Fig. 4.12. Possible dimeric components of MDI.

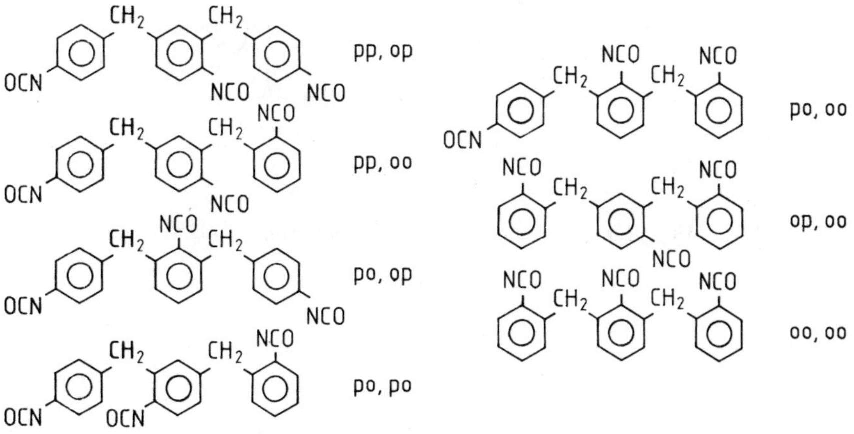

Fig. 4.13. Possible trimeric components of MDI.

An initial example of the use of MaxEnt in chemical analysis by NMR is provided by diphenylmethane diisocyanates (MDI) which are employed in the manufacture of a range of polymers. Synthesis of MDI leads to a complex mixture of dimers and higher oligomers. Figs 4.12 and 4.13 show all possible dimeric and trimeric products. The major product is the $4,4'$ isomer (denoted (para, para) or pp-). A reliable method for analysis of the mixture is required prior to polymerization. A typical MDI ^{13}C NMR spectrum is shown in Fig. 4.14. Although the pp-dimer can readily be identified, the other features, due to op- (ortho, para), and oo- (ortho, ortho) components cannot be assigned unequivocally. A further complication is that the relatively low concentration of the minor components results in weak spectral features, as illustrated by the methylene bridge regions for three different MDI samples (A, B and C) shown in Fig. 4.15. It is not

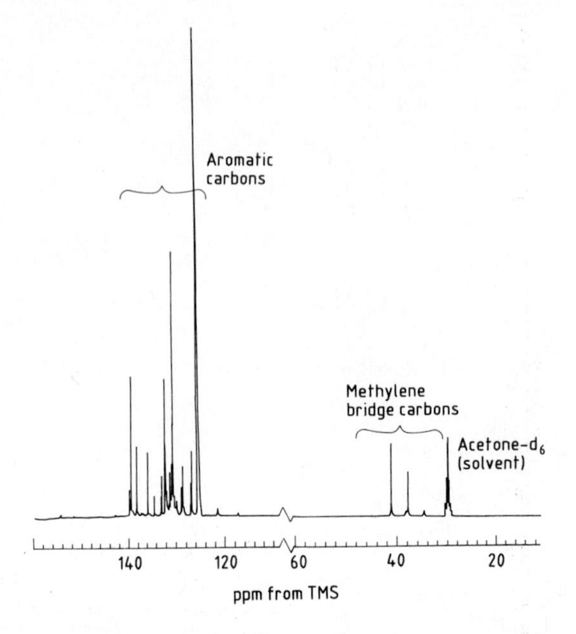

Fig. 4.14. A typical ^{13}C NMR spectrum of MDI.

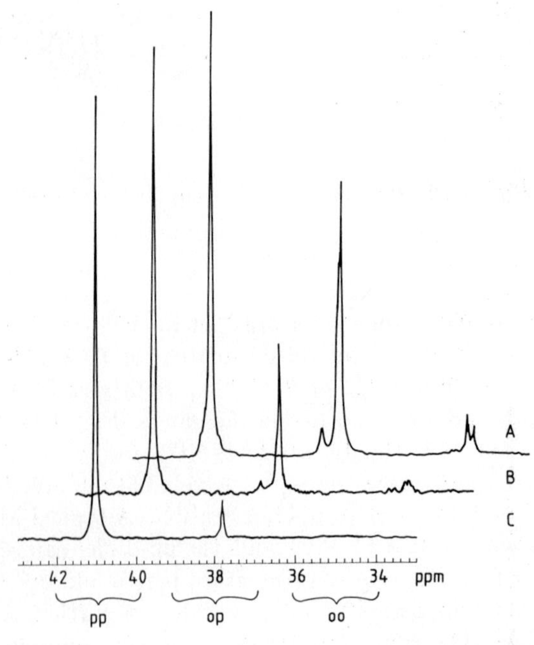

Fig. 4.15. The methylene bridge regions of MDIs A, B and C.

clear from the raw data whether the op- and oo- bridge concentrations are similarly distributed between dimers and trimers, nor is it clear from the spectra in Fig. 4.15 which are the predominant oo- components. Assignment of the methylene region would greatly facilitate the interpretation of the crowded aromatic region. The methylene region can be further subdivided into three areas:

1. 40–44 ppm pp-CH_2 bridges $(4,4')$,
2. 37–39 ppm op-CH_2 bridges $(2,4')$ and
3. 34–36 ppm oo-CH_2 bridges $(2,2')$.

The obvious choice for the model lineshape is to use the pp- feature which is expected to resemble closely the form of the other features in the spectrum. There is also good reason to expect that all pp- resonances will fall under this profile because of the limited influence of neighbouring bridged rings on the chemical shift of the pp- bridging methylene resonance. Fig. 4.16 shows MaxEnt reconstructions of the methylene region of the three MDI spectra from samples A, B and C. The MaxEnt analysis reveals a number of additional signals attributable to op- and oo- resonances which were not previously detected because of severe overlap and/or poor signal-to-noise ratio.

The validity of the assumption that the pp- resonance is a singlet was also tested by MaxEnt analysis with a calculated Lorentzian lineshape which revealed a sharp singlet, suggesting, as expected, that the substituent chemical shifts of trimeric (and higher) species are negligibly small.

Fig. 4.17 shows similar spectra and reconstructions for three different samples (D, E and F) which are pre-polymers (MDIs which have been partially reacted with a polyol). The MaxEnt analysis allows almost complete assignment of the spectra. From Fig. 4.16, C is seen to be the simplest material, containing only pp- and op-dimers. The op- peak is assigned to dimer rather than trimer or higher, as the sample is most unlikely to contain only trimers and no dimers. These assignments are then used to interpret the remaining MDI spectra. B contains only a small amount of the op-dimer, the remainder of the sample being trimeric or higher oligomers (it is not possible to distinguish between trimeric and higher species from the ^{13}C data, although higher oligomers are unlikely to be present at a significant level). The other major peak in the op- region of samples A and B is assigned to the pp-, op-trimer. Resonances characteristic of this structure are observed in the aromatic region of the spectrum. A number of additional resonances are observed in the op- and oo-methylene bridge regions and in the aromatic part of the spectrum. Fig. 4.18 shows the aromatic regions of spectra from samples A, C and F. Some tentative assignments have been made, as shown in Table 4.2.

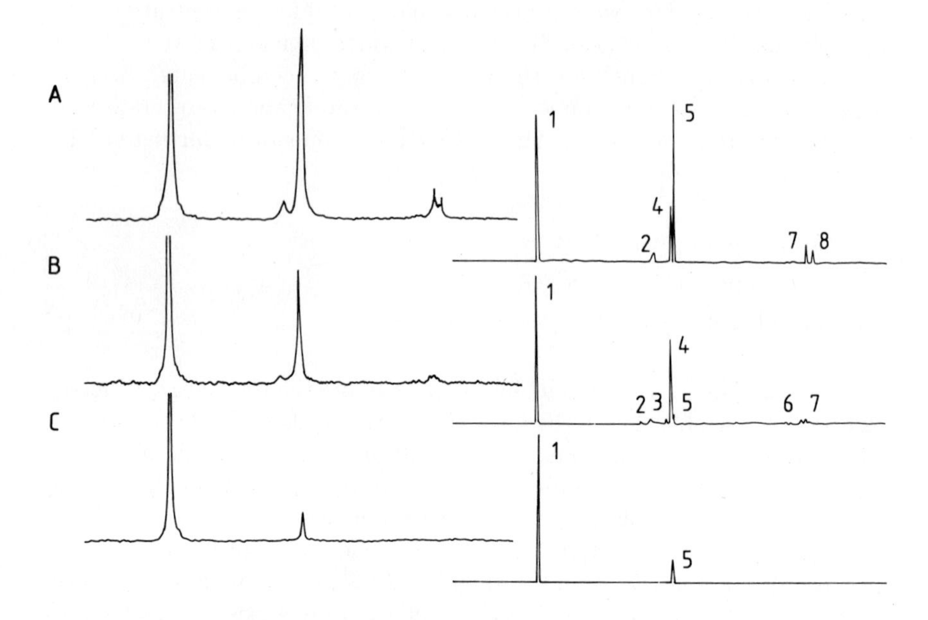

Fig. 4.16. MaxEnt reconstructions (right) from the data (left) of Fig. 4.15, using the pp- resonance as the blurring function.

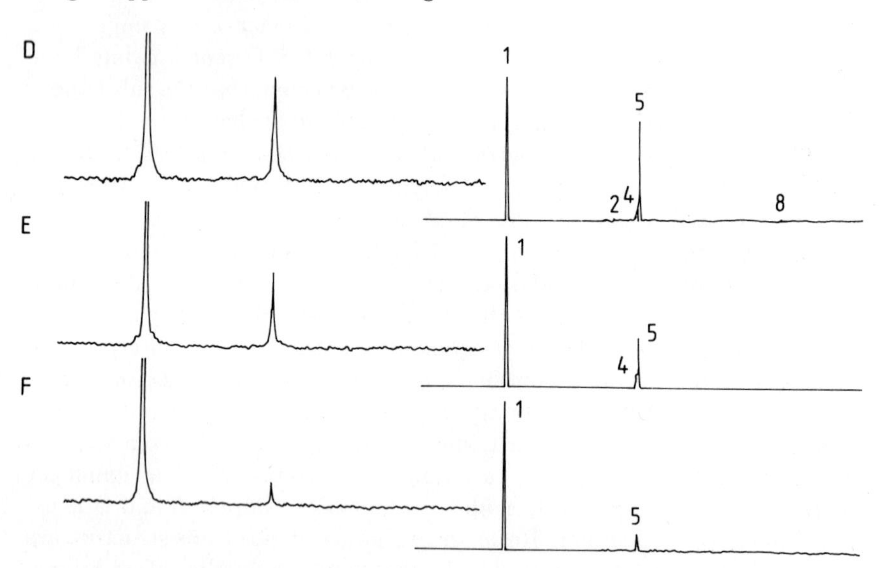

Fig. 4.17. The methylene bridge regions of the pre-polymers D, E and F (left) and their MaxEnt reconstructions (right).

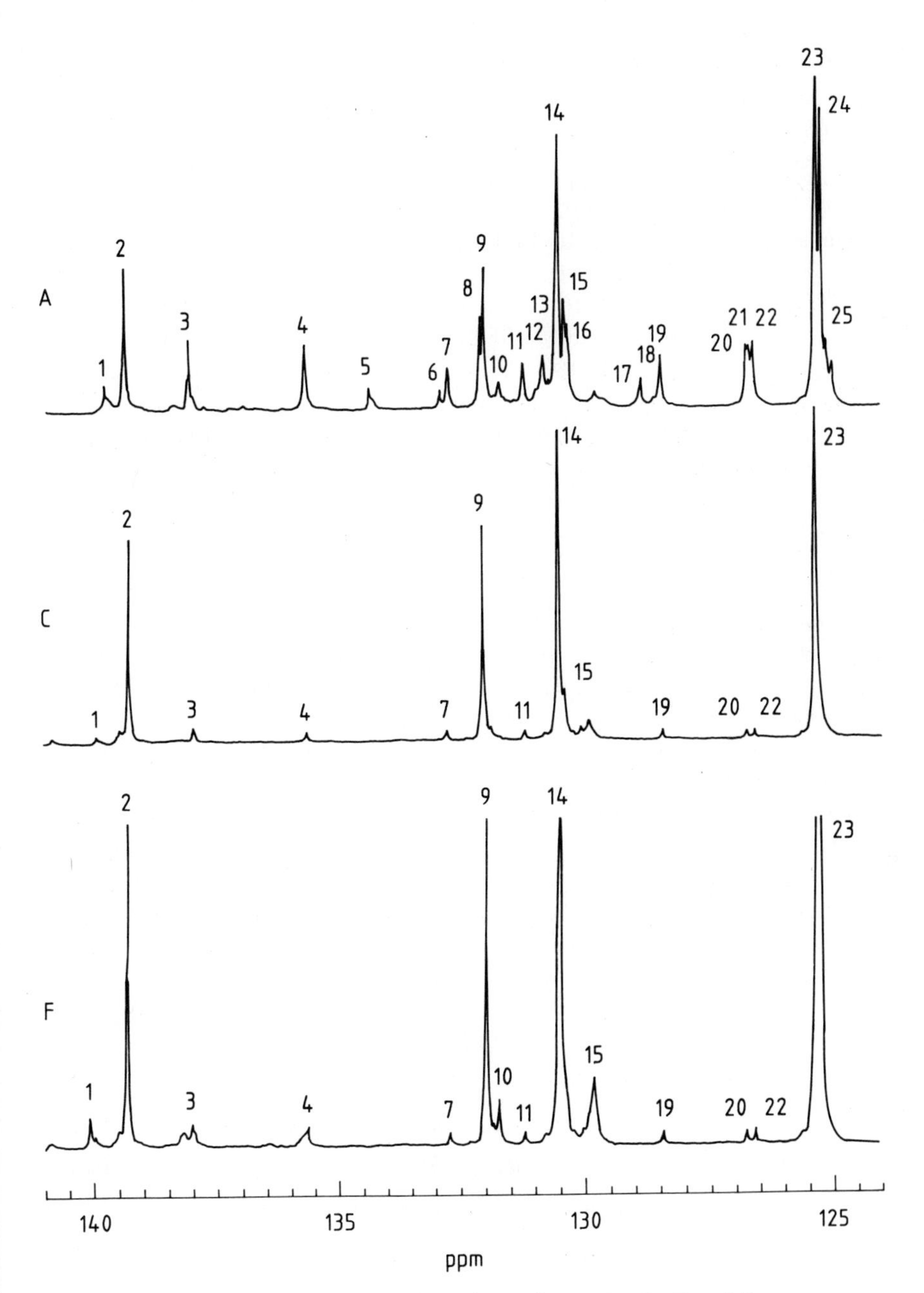

Fig. 4.18. The aromatic regions of samples A, C and F.

Table 4.2. ^{13}C chemical shifts of the aromatic carbons of some MDIs.

Compound	Ring	Aromatic C					
		1	2	3	4	5	6
Dimer 4, 4′	*A*	132.02	125.39	130.54	139.38	130.54	125.39
	B	132.02	125.39	130.54	139.38	130.54	125.39
Dimer 2, 4′	*A*	132.10	135.67	131.25	126.63	128.48	126.80
	B	132.75	125.30	130.42	138.05	130.42	125.30
Trimer	*A*	132.02	125.39	130.54	139.38	130.54	125.39
pp,op	*B*	130.83	135.67	131.73	139.77	128.89	126.73
	C	132.10	125.05	130.35	138.05	130.35	125.05

Notes: (*a*) The chemical shifts are referenced to DMSO-d6 solvent at 39.5 ppm w.r.t. TMS; (*b*) the phenol rings are designated *A*, *B* or *C* according to their position from the left in Figs. 4.12 and 4.13; and (*c*) the carbon numbering of each ring starts from the NCO-bearing carbon and proceeds around the ring in the direction of the nearest substituent.

The three pre-polymers have less complex structures as suggested by the spectra of Fig. 4.17. F contains only op- and pp-dimers. In addition to the dimers, D and E also contain some pp, op-trimer, the presence of which is confirmed by resonances in the aromatic region.

The example above is drawn from liquid phase NMR. The rapid motion of molecules within liquids averages to zero the strong internuclear dipole–dipole interactions, so liquid-phase NMR spectra are usually characterized by very sharp lines. By contrast, NMR spectra of the solid state tend to have very broad lines owing to little or no averaging of the strong internuclear interactions. The Hamiltonian for the direct dipolar interaction between two nuclei of spin $\boldsymbol{I}_1$ and $\boldsymbol{I}_2$ may be written in the form $A+B+C+D+E+F$, where

$$A = +(1-3\cos^2\theta)(\hat{I}_{z1}\hat{I}_{z2}), \tag{4.14}$$

$$B = -(1-3\cos^2\theta)(\hat{I}_1^+\hat{I}_2^- + \hat{I}_1^-\hat{I}_2^+)/4, \tag{4.15}$$

$$C = -3\sin\theta\cos\theta\exp(-\mathrm{i}\phi)(\hat{I}_{z1}\hat{I}_2^+ + \hat{I}_1^+\hat{I}_{z2})/2, \tag{4.16}$$

$$D = -3\sin\theta\cos\theta\exp(\mathrm{i}\phi)(\hat{I}_{z1}\hat{I}_2^- + \hat{I}_1^-\hat{I}_{z2})/2, \tag{4.17}$$

$$E = -3\sin^2\theta\exp(-2\mathrm{i}\phi)(\hat{I}_1^+\hat{I}_2^+)/4, \tag{4.18}$$

$$F = -3\sin^2\theta\exp(2\mathrm{i}\phi)(\hat{I}_1^-\hat{I}_2^-)/4, \tag{4.19}$$

where θ is the angle between the internuclear vector and the applied magnetic field and $\hat{I}_z$, $\hat{I}^+$ and $\hat{I}^-$ are the standard angular momentum operators. The two most important terms in the interaction are A and B which affect the energy of the nuclear spin states. Both terms have a $3\cos^2\theta - 1$

dependence. Thus, the resonant frequency of such a system is a function of the orientation of the internuclear vector relative to the applied field. If the angle between the internuclear magnetic field and the applied field is such that $3\cos^2\theta - 1$ is zero (the 'magic angle'), the dipolar splitting is also zero. Most samples are polycrystalline, having all possible orientations of the internuclear vector leading to broad lines. However, it is possible to make the internuclear vectors achieve an orientation where the average value of $3\cos^2\theta - 1$ is zero by spinning the sample very rapidly about an axis aligned at the magic angle relative to the applied field (the magic angle is the angle between the face of a cube and the body diagonal). The speed of rotation must be fast compared with the strength of the dipolar interaction for this technique to be fully effective.

The chemical shift is also a tensor quantity and is usually anisotropic, also giving rise to broad lines in powders. Fortunately, the nature of the anisotropy is rather similar to a dipolar interaction and is also averaged by the process of 'magic-angle spinning' (MAS).

Solid-state NMR spectra acquired with MAS often exhibit surprisingly sharp lines, allowing a good deal of useful chemical information to be obtained. Fig. 4.19 shows a ^{29}Si MAS spectrum and the MaxEnt reconstruction of a highly crystalline sample of the zeolite silicalite (recorded by Dr C. A. Fyfe). The multiplicity of lines is interpreted as arising from the existence of 24 non-equivalent crystallographic sites for silicon in the unit cell. The highest frequency peak (frequency conventionally increases to the left in NMR spectra) is known to correspond to a single site, and is used as the model lineshape in the deconvolution process.

A number of interesting points arise from a comparison of Fig. 4.19*a* and *b*. The raw data has eighteen distinguishable maxima, and a 'shoulder' discernible on the low frequency side of the largest spectral feature. Fig. 4.19*b* has twenty maxima, with most lines resolved down to the baseline level. In particular, the group of features marked B, comprising a broad line flanked by a shoulder to low frequency and a barely resolved peak to high frequency, are revealed as a group of four lines whose integrals accurately reflect the total intensity of this feature (six), relative to those lines known to be of unit intensity.

A second point arises from examination of the group of lines marked A. The overall intensity of this group is, as required, five. However, the integrals for the original lines in the group are close to $\frac{4}{3} : \frac{5}{3} : 2$ instead of the expected $1 : 2 : 2$ based on a 24-site structure. The origin of this effect is not clear. Further investigation would be required in order to ascertain whether the effect is a genuine feature of the silicalite spectrum or some artefact of the reconstruction process.

A further feature is that the lines do not, of course, all have the same width. This is clear in the reconstruction for well-resolved lines representing single sites which do not all have the same height. The effect is particularly

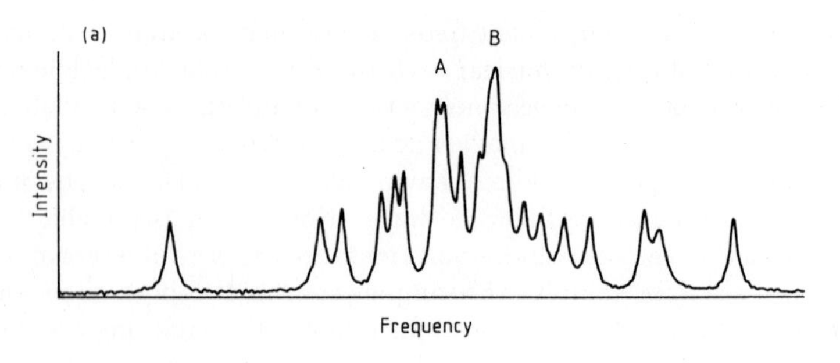

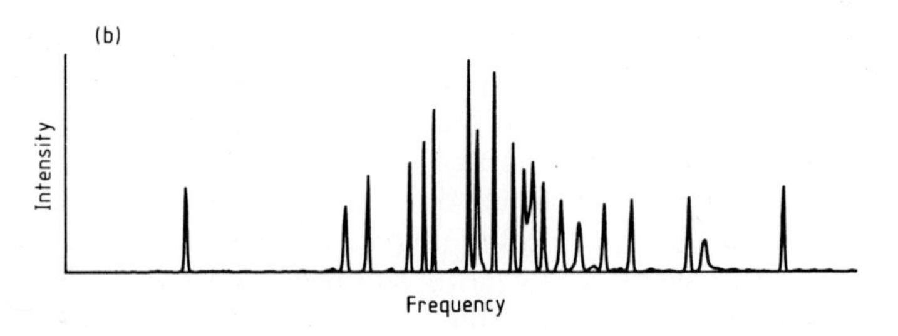

Fig. 4.19. (*a*) ^{29}Si MAS spectrum of the zeolite silicalite. (*b*) The MaxEnt reconstruction using the lineshape of the peak of highest frequency in *a*.

pronounced for the second peak at the low-frequency end of the spectrum. Thus, the lineshapes of these lines are not perfectly described by the experimental profile used in the reconstruction. Nevertheless, the integrals under these lineshapes have the correct value, illustrating the tolerance of the method to the choice of blurring function. However, if an experimental lineshape differs significantly from the chosen lineshape, errors in intensity and even false splittings may occur in extreme cases. This should be borne in mind when selecting the model lineshape.

Despite the success of MaxEnt in the deconvolution of NMR spectra, the fact cannot be ignored that the user has to impose some bias on the data prior to MaxEnt processing, in the implementation described above. Many NMR experiments are considerably more complicated than those described above and yield spectra where the signals have widely differing phase. Whilst there have been attempts to apply MaxEnt to such spectra by allowing each type of signal to form a separate distribution (Laue *et al.* 1985), we find this approach entirely unconvincing. In our opinion, the necessity for user intervention in this implementation of MaxEnt undermines the status of MaxEnt as the philosophically 'correct' method of data processing.

A number of ideas to circumvent the problems of applying MaxEnt to phase coherent data have been proposed. The most promising approach to date appears to be that of Hore and Daniell (discussed in Chapter 3) where the conventional entropy expression is applied to the initial spin populations in the classical case (populations are, of necessity, real and positive), or the spin density operator in the quantum-mechanical case. The principal attraction of this approach is that it builds the physics of the NMR experiment into the model, thereby maximizing the use of prior information. An earlier suggestion made by Wright and Belton (1986) was to use an alternative definition of the entropy involving the square moduli of the intensities, thereby removing the phase problem. The Wright and Belton entropy expression is drawn by analogy with the von Neumann equation for the entropy of an ensemble of isolated nuclear spins. The entropy expression thus has no firm theoretical basis. Since Wright and Belton presented no experimental results, it is a useful exercise to see how well their method works in practice.

4.5 The Wright–Belton approach

4.5.1 Mathematical outline

Wright and Belton's entropy definition S', say, in square modulus form, for intensity $\mathbf{\Phi}$ is

$$S' = -\sum_j q_j \ln q_j, \tag{4.20}$$

where $q_j = |\Phi_j|^2 / \sum_j |\Phi_j|^2$. Using this definition, a phase-free approach to MaxEnt will require a constraint which is independent of the phase of the data. The chi-squared function defined in (4.11) is not appropriate for this purpose, as it requires the mock data F_i to be correctly phased to match the raw data D_i. Otherwise, incorrect phasing will give rise to biased chi-squared which will misdirect the algorithm.

To reconstruct a phase-free data constraint, we shall start with a model for the complex FT-NMR data D_t acquired in quadrature (dual channel) mode, whose real and imaginary parts correspond to components of signal in phase and 90° out of phase with respect to the radiofrequency carrier signal. The model will be written as the sum of signal Z_t and noise ϵ_t, $D_t = Z_t + \epsilon_t$, where (using $\sum_\nu$ for summation over ν from 0 to $N-1$, N being the sample size)

$$Z_t = \exp\big(-\mathrm{i}\psi - K(t+t_0)\big) \sum_\nu f_\nu \exp\big(-2\pi \mathrm{i}\nu(t+t_0)/N\big). \tag{4.21}$$

Here f_ν is the spectrum, ψ is the phase, t_0 is the time delay and K $(= 1/T_2)$ is the decay rate. The real and imaginary parts of ϵ_t will be assumed to be independent Gaussian, each with expectation zero and variance σ^2. The noise is further assumed to be uncorrelated with the signal.

If the spectrum is phased by multiplying f_ν by $\exp(-\mathrm{i}\psi - 2\pi\mathrm{i}\nu t_0/N)$ and the phased spectrum is denoted by f'_ν, then for $\Phi_\nu = f'_\nu \exp(-Kt_0)$, (4.21) can be rewritten as

$$Z_t = \exp(-Kt) \sum_\nu \Phi_\nu \exp(-2\pi\mathrm{i}\nu t/N). \tag{4.22}$$

Here Φ_ν is the phased spectrum, reduced in intensity by $\exp(-Kt_0)$, which will be identified with Wright–Belton's Φ_j in S'. A solution for $|\Phi_\nu|$ will provide a solution for $|f_\nu|$ when t_0 is known or estimated by other means (for example, post-MaxEnt application of the maximum likelihood method).

The logical basis for a phase-free consistency constraint is the auto-correlation function of the raw data, which is closely related to the Fourier transform of $|\Phi_\nu|^2$. (Auto-correlations provide measures of lag correlations between observations in a time series.) For convenience of computation we shall use the auto-covariance function which is an unscaled version of the auto-correlation. For a given lag s, the auto-covariance function of a complex series X_t is defined as the expectation of the product of lagged observations $X_t^* X_{t+s}$ over time t. Similarly, the cross-covariance of two complex series X_t and Y_t is defined by the average of $X_t^* Y_{t+s}$. We shall use argument s for lag s, $\gamma_{XY}(\cdots)$ for an ensemble average and $C_{XY}(\cdots)$ for its finite sample estimate defined by

$$C_{XY}(s) = \sum\nolimits^s (X_t^* Y_{t+s})/N, \tag{4.23}$$

where $\sum^s$ stands for summation over time t from 0 to $N - s - 1$.

The function $C_{DD}(s)$ calculated for $s = 0, 1, 2, \ldots,$ will form a reconstructed time series upon which subsequent data processing can be based. Since $D_t = Z_t + \epsilon_t$, it can be seen that

$$C_{DD}(s) = C_{ZZ}(s) + e(s). \tag{4.24}$$

Here $e(s)$ is the error term arising from the cross-products of signal-noise and noise-noise combinations. As Z_t and ϵ_t are uncorrelated, the expected value of $C_{DD}(s)$, $m(s)$ say, satisfies

$$m(s) = \gamma_{ZZ}(s) + 2\sigma^2\delta(s), \tag{4.25}$$

where $\gamma_{ZZ}(s)$ is the auto-covariance function of Z_t, given by

$$\gamma_{ZZ}(s) = g(K)\exp(-Ks) \sum_j |\Phi_j|^2 \exp(-2\pi\mathrm{i}js/N), \tag{4.26}$$

for $g(K) = \sum_s \exp(-2Kt)$ and $\delta(s)$ a delta function. Equation (4.26) is the damped Fourier transform of $|\Phi_j|^2$ which is very similar to (4.22). This equation clearly suggests that the inverse FT of $C_{DD}(s)$ will exhibit Lorentzian lineshapes in the domain of $|\Phi_j|^2$.

For a MaxEnt-based phase-free formulation using S', we may now formulate a chi-squared function C' in terms of $|\Phi_\nu|^2$ by using the reconstructed series formed by $C_{DD}(s)$. The problem will then be one of maximizing a function $Q' = S' - \lambda C'$, say, to find a solution for $|\Phi_\nu|^2$ by using MEMSYS.

The chi-squared function of (4.11) in square-additive form is based on independent Gaussian noise ϵ_t. The assumption of independence does not apply to $e(s)$. As $e(s)$ is composed of signal-noise cross-products, it is signal dependent and its statistical properties are different from those of ϵ_t. The elements of the error sequence $e(0)$, $e(1)$, $e(2)$, ..., will have overlapping signals giving rise to correlations. This calls for a more broadly based formulation of the chi-squared function, that applies to correlated variates. The new formulation will be an extended version of (4.11) involving the square and product terms of mean-subtracted auto-covariances (appropriately weighted).

The variance–covariance or 'dispersion' matrix $\mathbf{V}$, say, of the reconstructed series $C_{DD}(s)$ is very complicated. The (s,r)th element V_{sr} of $\mathbf{V}$ is the covariance of $C^*_{DD}(r)$ and $C_{DD}(s)$, which equals

$$\begin{aligned} &N^{-2}\exp(-K(r+s))\sum_{p,q}\left(|\Phi_p|^2|\Phi_q|^2\exp(-2\pi iq(s-r)/N)H(p,q)\right) \\ &+N^{-1}2\sigma^2 g(K)\rho(s-r)\exp(-Ks)2\sinh(Kr) \\ &+N^{-1}4\sigma^4\delta(r)\delta(s)+N^{-1}(1-s/N)4\sigma^4\delta(s-r), \end{aligned} \tag{4.27}$$

where $\rho(\cdots)$ is the discrete Fourier transform of $|\Phi_\nu|^2$ and $H(p,q) = \left((1-b)^2 + 4b\sin^2[\pi(p-q)/N]\right)^{-1}$ with $b = \exp(-2K)$. It may be verified that V_{rs} is the complex conjugate of V_{sr}. Hence $\mathbf{V}$ is complex Hermitian. The diagonal elements of $\mathbf{V}$ represent variances.

The dispersion matrix $\mathbf{V}$ has some important features. The off-diagonal elements (V_{rs} with $r \neq s$) diminish exponentially with increasing row and column indices, r and s. The diagonal elements also diminish down the matrix to a constant value (the last term).

The joint probability distribution of the auto-covariances, upon which the chi-squared function is to be based, is also extremely complicated. Nevertheless, the general nature of the distribution is asymptotically multivariate Gaussian. Let $\boldsymbol{C}$ be the vector of auto-covariances up to a maximum lag k and $\boldsymbol{M}$ be its expectation vector given by (4.25). Then the multivariate formulation of the chi-squared function is

$$C' = (\boldsymbol{C}-\boldsymbol{M})^*\mathbf{V}^{-1}(\boldsymbol{C}-\boldsymbol{M}) \tag{4.28}$$

and the constraint to be satisfied is $C' = C_{\text{aim}}(= k)$.

For rapid decay rates (K moderate to large) the off-diagonal elements of $\mathbf{V}$ will quickly diminish and the diagonal elements will dominate. In that case the chi-squared function can be approximated by a simpler function, similar to (4.11). The 'truncated' (diagonal) form of chi-squared is used in the work described below to enable the Wright–Belton approach to be implemented using the `MEMSYS` search routines.

4.5.2 Results

The utility of the Wright–Belton approach was tested using some simple simulated free induction decay data. The first test of the method was upon phased data consisting of three signals with relative intensities 1 : 2 : 2 to which Gaussian noise was added. The spectrum corresponding to this data is shown in Fig. 4.20*a*. The reconstruction obtained with the Wright–Belton entropy definition is shown in Fig. 4.20*b*. The reconstruction is clearly most unsatisfactory, in terms of both the degree of noise suppression and the residual wiggles around the signals. The quality of the reconstruction can be greatly improved by the introduction of an arbitrary baseline parameter b (Hore 1985)

$$S' = -\sum_j q_j \ln \frac{q_j}{b}. \tag{4.29}$$

The parameter b may be thought of as a means of incorporating prior knowledge into the reconstruction; a low value of b forces the baseline towards zero in the absence of any signal (where we know that it should be). Fig. 4.21 shows a reconstruction of the data of Fig. 4.20 obtained with this modified definition of the Wright–Belton entropy. It is quite acceptable, at least in a qualitative sense.

The modified entropy definition was then tested upon data of varying phase. Fig. 4.22 shows the real parts of two spectra, where the three signals have relative phase 0°, 45° and 90° (from the left), and the relative intensities are 1 : 2 : 2 in Fig. 4.22*a* and 2 : 1 : 2 in Fig. 4.22*b*. The reconstructions of these spectra are shown in Fig. 4.23. Fig. 4.23*a* is a defensible reconstruction of the data, but the central peak is absent in Fig. 4.23*b*. This shows that the reconstruction is not always reliable.

Further investigation shows that the relative intensities of the lines in the reconstructions are strongly dependent upon the degree of deconvolution attempted. This point is illustrated in Fig. 4.24 which shows a comparison of the relative intensities obtained using `MEMSYS` with the conventional entropy definition, and the Wright–Belton entropy, after reconstruction of data from in-phase signals with relative intensity 2 : 1 : 2. The inescapable conclusion is that this implementation of the Wright–Belton entropy does

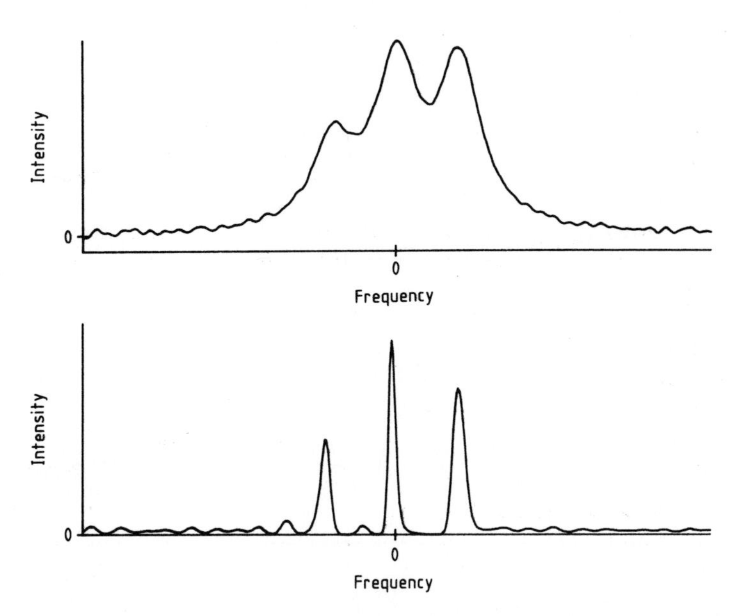

Fig. 4.20. (a) In-phase test data (relative signal intensities 1 : 2 : 2). (b) Reconstruction using the simple Wright–Belton entropy definition.

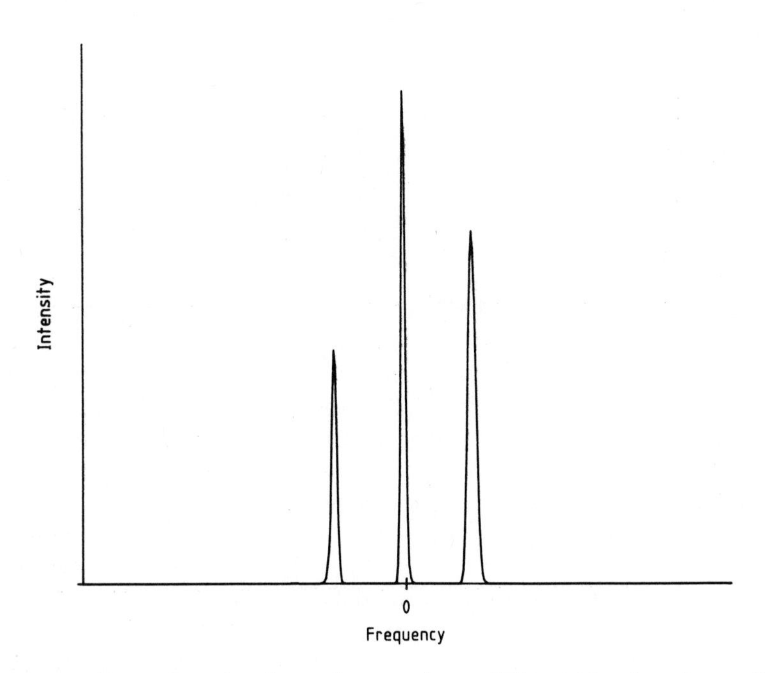

Fig. 4.21. Reconstruction from the test data of Fig. 4.20 using the modified Wright–Belton entropy.

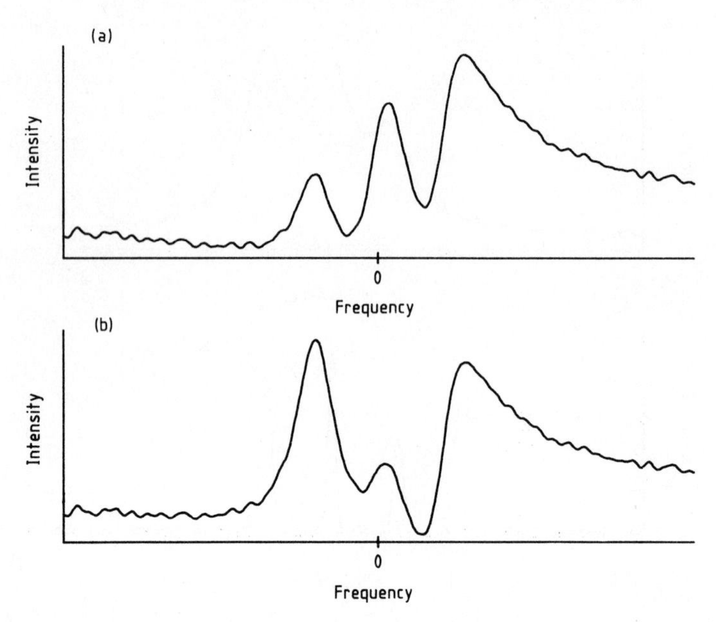

Fig. 4.22. Out-of-phase data (0° : 45° : 90°) with relative intensities (*a*) 1 : 2 : 2 and (*b*) 2 : 1 : 2.

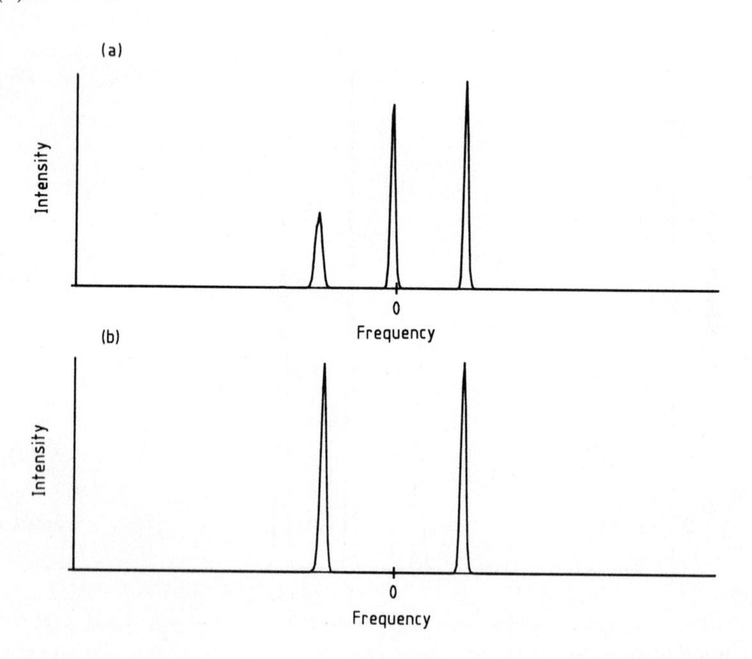

Fig. 4.23. Wright–Belton reconstructions from the data of Fig. 4.22.

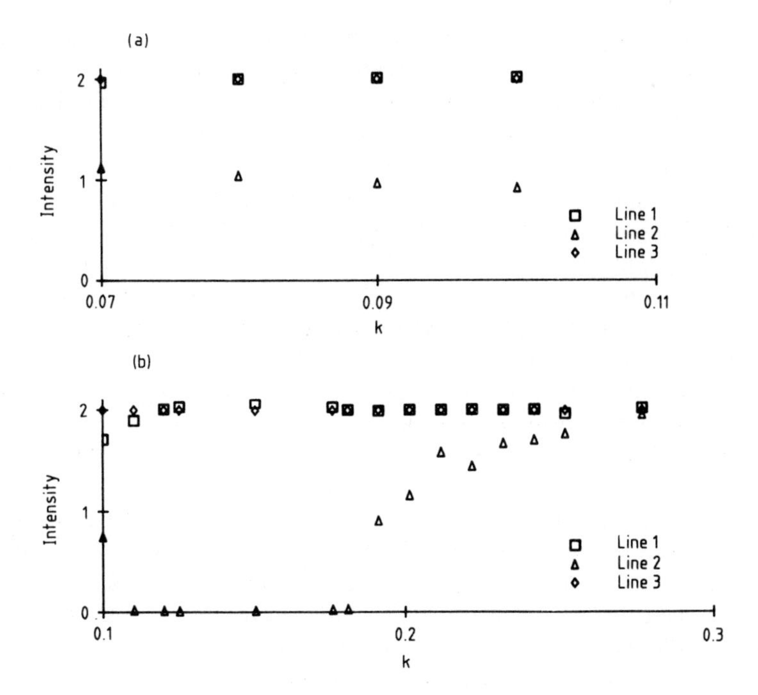

Fig. 4.24. Relative intensities (*a*) in normal `MEMSYS` and (*b*) in Wright–Belton reconstructions, calculated from in-phase signals with relative intensity 2 : 1 : 2. The k factors used differ because of the nature of $m(s)$—see (4.25) and (4.26).

not yield quantitative results, and is therefore unsuitable for analytical spectroscopy. Nevertheless, it should be borne in mind that an approximate form for the chi-squared has been used to enable use of the `MEMSYS` package. An exhaustive investigation of the Wright–Belton approach would require the development of modified search algorithms to enable the full model to be tested.

4.6 Conclusions

The examples discussed above show that MaxEnt is a very powerful data processing technique, capable of producing some astonishing results, and as such is a most useful tool for the interpretation of 'difficult' spectroscopic data. However, despite the fact that the algorithm is relatively tolerant of the choice of blurring function, the best results will only be obtained when an informed choice is made, based upon a sound knowledge of the design of the instrument and the experimental conditions. Therefore, blanket use of the method for processing routine spectroscopic data is definitely not recommended. Similarly, sound spectroscopic judgement must be applied to the interpretation of MaxEnt reconstructions which must be defensible in terms of the physics and chemistry of the problem.

The most appealing feature of MaxEnt is that, subject to a realistic choice of blurring function, the reconstruction will not contain any response for which there is no evidence in the original data. The corollary is that the absence of a 'known' spectral feature in a reconstruction shows that there is no evidence for it in the raw data; that is, the raw data is of insufficient quality. Thus, in principle, MaxEnt may also be used as a criterion for the adequacy of experimental data. This sort of analysis could be used to advantage in defining some optimum strategy (for example, in terms of signal-to-noise or limiting resolution) for a set of experiments upon a series of samples where there are only minor changes in composition between the samples.

There is clearly much work to be done even for standard applications (that is, those which are adequately described by positive additive distributions) to enable reliable error bars to be placed upon the image (see the discussion in Chapter 2). The development of applications of MaxEnt to phase-coherent data remains a problem, but the approach of Daniell and Hore (see Chapter 3) appears promising. MaxEnt appears to have a bright future in many areas of spectroscopy, particularly given the rapid proliferation of fast and relatively inexpensive computer hardware.

Acknowledgements

We wish to thank Mr C. J. Dyos, Dr J. D. Boyle, and Dr K. P. J. Williams for their contributions to this work. Dr C. A. Fyfe is thanked for providing the silicalite NMR data.

We are grateful to BP Research for permission to publish this work.

References

Grant, A. I. and Packer, K. J. (1989). Enhanced information recovery from spectroscopic data using MaxEnt. In *Maximum entropy and Bayesian methods, Cambridge, England, 1988* (ed. J. Skilling), pp. 251–9. Kluwer, Dordrecht.

Hore, P. J. (1985). NMR data processing using the maximum entropy method. *Journal of Magnetic Resonance*, **62**, 561–7.

Laue, E. D., Skilling, J., and Staunton, J. (1985). Maximum entropy reconstruction of spectra containing antiphase peaks. *Journal of Magnetic Resonance*, **63**, 418–24.

Loader, J. (1970). *Basic laser Raman spectroscopy.* Heyden, London.

Skilling, J. (1984). *MEM Users' Guide.* Maximum Entropy Data Consultants Ltd, 33 North End, Meldreth, Royston, SG8 6NR.

Skilling, J. and Gull, S. F. (1985). Algorithms and applications. In *Maximum entropy and Bayesian methods in inverse problems* (ed. C. R. Smith and W. T. Grandy), pp. 83–132. Reidel, Dordrecht.

Stephenson, D. S. (1988). Linear prediction and maximum entropy methods in NMR spectroscopy. *Progress in Nuclear Magnetic Resonance Spectroscopy*, **20**, 515–626.

Wright, K. M. and Belton, P. S. (1986). A new definition of the information content of N.M.R. spectra suitable for use in maximum entropy signal processing. *Molecular Physics*, **58**, 485–95.

5

Maximum entropy and plasma physics

G. A. Cottrell

Abstract

Experiments on magnetically confined fusion plasmas, for example the Joint European Torus (JET) tokamak, require a range of diagnostic techniques for measurement of plasma quantities such as temperature and density. Usually, some kind of inverse transform is needed to convert measured raw signals into useful data. Also, there are often problems arising from inadequately sampled data and instrumental resolution limitations. It will be explained how the maximum entropy method (MaxEnt) can be applied in such cases with three advantages over conventional analysis: the generation of positive solutions, the suppression of noise and the suppression of spurious transform-related artefacts.

MaxEnt is illustrated using a number of examples from the field, involving: the Fourier transform (for example, Michelson interferometry), deconvolution of spectra, and two-dimensional tomography of atomic beams. The advantages of MaxEnt are seen clearly in the problem of extracting the maximum amount of information, for example on plasma profiles, starting from only a limited and noisy data set. Used in this way, the method responds flexibly to the demands of the data and does not suppress any significant features.

5.1 Introduction

Much of the driving force behind laboratory plasma physics research comes from the prospect of using controlled thermonuclear fusion, based on either inertial or magnetic confinement, to provide economically significant amounts of power. A magnetically confined high-temperature plasma is highly complex and, to make progress in our understanding of the fundamental plasma physics, we must be able to make accurate measurements of internal plasma parameters. Only then can theoretical ideas be confronted quantitatively with experiment. The specific goals of fusion impose stringent constraints on temperature, density and confinement and so give rise to extra problems for plasma diagnosis. For example, the high central temperatures of thermonuclear plasmas exclude direct measurements using the intrusive material probes which are used mainly to monitor the cool exterior plasma regions. A wide range of non-intrusive physical measurement techniques has therefore been developed for diagnosis, ranging from basic measurements with electromagnetic probes and spectroscopy to nuclear particle and photon measurements. Most methods involve gathering data which has been transformed in some way either by the data collection geometry or by instrumental limitation and so inverse transformation is needed to obtain local plasma physics quantities.

In many measurements, the desirable high degree of line of sight access to the plasma has to be balanced against technological constraints; access is often restricted by the placement of magnetic field coils or other structures. In future reactor-like fusion devices, diagnostic access is likely to be further limited by the need to include first-wall and reactor blanket systems. Additional limitations arise in the measuring instruments themselves; typical examples are the finite resolution of spectrometers and the truncation of the measured autocorrelation functions in Fourier transform spectroscopy. In these cases we are dealing with the problem of sparse data, where the information is also corrupted by the inevitable noise. The use of linear inversion methods in this context can exaggerate defects in the data, in some cases leading to the appearance of unwanted transform-related artefacts in the reconstruction and in the worst cases giving misleading results.

Here we consider the use of a radically different scheme for analysing difficult inverse problems. It is based on the idea that the unknown quantities that we wish to reconstruct from the data can be described in terms of positive, additive distributions. We note that a large number of plasma physics measurements belong to this category, for example, the number of particles per unit volume or the number of photons per unit wavelength. We ask: what is the most probable configurational arrangement of particles or photons in their respective cells of volume or wavelength interval that is consistent with the data that we have actually measured? The question is answered by maximizing the configurational entropy of the reconstruction subject to the constraint that the reconstruction be consistent with the measured data and its experimental uncertainties. Four different physical measurements, each involving an inverse transform in the analysis of diagnostic data from research in controlled thermonuclear fusion, are described and analysed within the framework of the maximum entropy (MaxEnt) formalism. The generality of MaxEnt is stressed; it is also applicable to many similar data analysis problems which are encountered outside the specific field of plasma diagnosis. The four cases dealt with here are:

1. inversion of Abel transformed data,
2. deconvolution of electromagnetic line spectra,
3. Fourier transform spectroscopy, and
4. two-dimensional tomographic reconstruction with sparse data.

To illustrate some of the particular problems encountered in extracting information from laboratory plasmas, it is first necessary to give a brief overview of the essential features of experimental work in controlled magnetic fusion.

5.1.1 Overview of controlled thermonuclear fusion

When nuclei of the isotopes of hydrogen, deuterium, and tritium come sufficiently close to one another, short range nuclear forces fuse them together resulting in the release of an alpha particle and a neutron. The nuclear rearrangement results in a reduction of total mass and a release of binding energy which appears as kinetic energy of the reaction products:

$$\mathrm{D} + \mathrm{T} \rightarrow {}^{4}\mathrm{He}(3.5\,\mathrm{MeV}) + \mathrm{n}(14.1\,\mathrm{MeV}). \tag{5.1}$$

Before short-range nuclear forces can take effect, the mutual electrical repulsion of the nuclei must be overcome. This means that fusion reaction cross-sections are extremely small unless the relative velocities of the nuclei are high enough to overcome their electrostatic potential energy. For nuclei having a Maxwellian velocity distribution, the high characteristic velocities needed to produce a high fusion reaction rate mean that the temperature must be high and that the particles should be in the fully ionized plasma state. To produce a significant reaction rate from (5.1), the plasma temperature T must be in excess of about $10\,\mathrm{keV}$ ($\equiv 10^8\,\mathrm{K}$). The possibility of using the DT reaction to form the basis of a fusion reactor is attractive since just 1 kg of fuel would release 10^8 kWh of energy and would provide the requirements of a 1 GW (electrical) power station for a day.

The present goal of research in controlled thermonuclear fusion is to produce and study plasmas in conditions approaching those anticipated in an ignited deuterium–tritium reactor. To obtain a net energy gain in this system, the thermonuclear power must exceed the continuous loss of energy from the plasma. Energy losses occur through various mechanisms of heat conduction and particle convection across the confining magnetic field as well as through bremsstrahlung radiation. A measure of the quality of thermal insulation in a reactor can be expressed in terms of the energy replacement time,

$$\tau_E = \frac{W}{P}, \tag{5.2}$$

where W is the total plasma energy and P the total power needed to sustain the plasma in a steady state. In a magnetic containment device, net energy gain from reaction (5.1) is given by the condition that the **Lawson product** of the particle number density n and the energy replacement time be greater than about $10^{20}\,\mathrm{m}^{-3}\mathrm{s}$. To achieve both this condition and that of high plasma temperature, the magnetic containment device known as a **tokamak** is widely considered to offer the most promise.

5.1.2 Overview of the tokamak

In the tokamak configuration (Wesson 1987), plasma is confined in the geometry of a torus and is symmetric about the major axis. Fig. 5.1 shows its main features. Characterizing the device is a strong toroidal magnetic field B_{T} generated by external coils, supplemented by a second, weaker, poloidal field component B_{P} produced by a large current I_{P} flowing in the plasma itself. In the Joint European Torus (JET) tokamak (Pease and Bickerton 1987), the toroidal field and plasma current can be as high as 3.5 tesla and 7 MA respectively. In addition, a third field in the vertical direction is needed to counteract the natural tendency of the plasma current ring to expand. The main toroidal field of the tokamak varies inversely with the major radial coordinate R of the torus as

$$B_{\mathrm{T}}(R) = \frac{B_{\mathrm{T}}(0)R_0}{R}, \tag{5.3}$$

where $B_{\mathrm{T}}(0)$ is the magnetic field strength measured at the major radius of the plasma centre R_0. The poloidal field, necessary for stability and equilibrium, combines vectorially with the toroidal field to produce a net helical field which winds around the torus. Without this helical twist, charge separation of particles gyrating about the field lines would set up a vertical electric field $\boldsymbol{E}$, leading to a rapid outward drift of the plasma in the $\boldsymbol{E} \times \boldsymbol{B}$ direction. The combination of toroidal and poloidal magnetic fields gives rise to a set of nested, closed surfaces of constant magnetic flux on which the plasma ions and electrons move. In directions parallel to the flux surfaces the particle and thermal diffusivities are many times larger than those in the perpendicular direction; thus inhomogeneities in plasma pressure within a flux surface are rapidly smoothed out. This leads to a pressure gradient in the radial direction.

Apart from its role in providing a stable equilibrium, the plasma electron current also heats the plasma through collisional Joule dissipation. However, with increasing electron temperature T_{e} the plasma becomes increasingly collisionless and the electrical resistivity decreases (as $T_{\mathrm{e}}^{-\frac{3}{2}}$) so that, in some devices, Joule heating alone is not sufficient to produce the high temperatures required for fusion. Some form of additional plasma heating is therefore required, of which there are two main sources. One

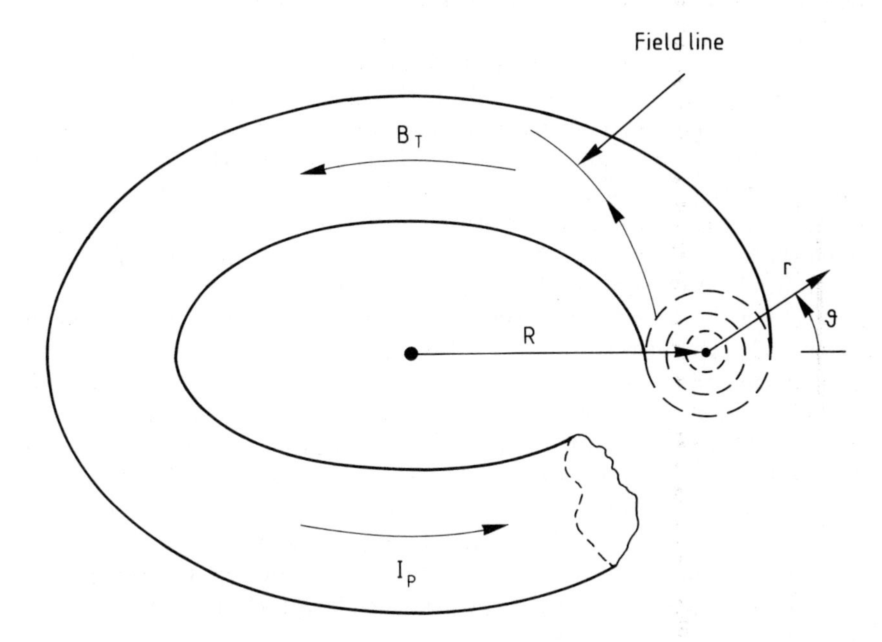

Fig. 5.1. The main features of a tokamak showing the directions of the toroidal field B_{T} and plasma current I_{P}. The major and minor radial coordinates are R and r. Shown by dashed lines are the locations of the magnetic flux surfaces. In the Joint European Torus (JET) tokamak, the major and minor radii are $R_0 = 2.96\,\mathrm{m}$ and $a = 1.2\,\mathrm{m}$ with the plasma shape being elongated in the vertical direction by a ratio of less than 1.65.

of these is the application of intense radiofrequency wave energy to the plasma which is absorbed directly by resonant particles at their respective gyrofrequencies in the magnetic field. In JET, approximately 20 MW of radiofrequency heating power has been applied to the plasma in the ion cyclotron (25–50 MHz) frequency range. The other is that of neutral beam injection (NBI). With this, a beam of energetic neutral atoms (with energies $E_{\mathrm{B}} \gg kT$) is injected through the confining tokamak magnetic field into the plasma. Once there, the fast atoms are collisionally ionized, become trapped in the magnetic field and transfer their energy to the plasma by Coulomb collisions. In JET, 16 individual neutral hydrogen or deuterium beams have been used to deliver up to 20 MW of power to the plasma. Both the radiofrequency and NBI methods are proven and effective methods of supplementing Joule heating and have been found experimentally to give similar plasma heating efficiencies.

5.2 Illustration of the maximum entropy method

To introduce and illustrate the method of maximum entropy (MaxEnt), a simple and commonly encountered problem has been chosen: inversion of Abel transformed data in tokamak measurements of the number density of electrons n_e. The number density of electrons clearly belongs to the class of positive, additive distributions. The ideas discussed are, naturally, relevant to the determination of other positive, additive distributions. The signal detected by a diagnostic is composed of a local quantity of interest which has been integrated along an instrumental line of sight through the plasma. In the present discussion, the quantity of interest is the number density of plasma electrons. In a tokamak plasma, n_e is a function of the minor radius r; the problem is to measure $n_e(r)$ given line of sight data for a number of different chords in the plasma. The information is used to infer the most likely radial distribution of n_e which gives rise to the measured data. Because electrons are assumed to move essentially freely on flux surfaces, their distribution can be considered to be tied to the underlying structure of the magnetic flux surfaces and the problem becomes one of determining the radial distribution. With this geometry, the problem of inverting integrated line of sight data can then be attacked using the inverse Abel transform.

5.2.1 Measurement of electron density

One of the most successful and accurate of the class of non-intrusive plasma diagnostics involves the use of electromagnetic waves as a probe. Low intensity waves cause negligible perturbation to the plasma but still allow information about the internal plasma properties to be gathered with good spatial resolution. Here we are concerned with the refractive properties of the plasma, that is, the effect that the dielectric properties of the plasma have on electromagnetic wave propagation.

In a hot plasma the refractive index is governed by the population of free electrons and is given by

$$\mu^2 = 1 - \frac{\omega_p^2}{\omega^2}, \tag{5.4}$$

where ω_p is the plasma frequency,

$$\omega_p = \left(\frac{n_e e^2}{\epsilon_0 m_e}\right)^{\frac{1}{2}}, \tag{5.5}$$

and e and m_e are, respectively, the electron charge and mass, and ϵ_0 is the permittivity of free space. The refractive index of the plasma is typically

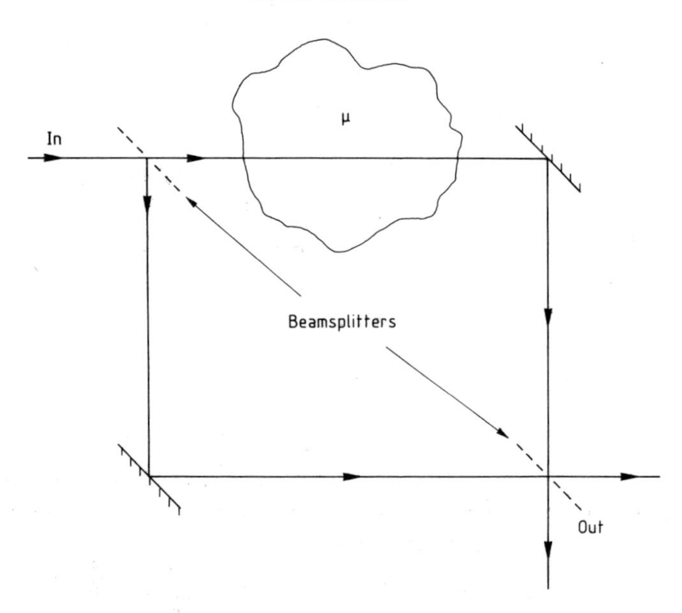

Fig. 5.2. Mach–Zhender interferometer configuration used to measure the refractive index μ of the plasma.

measured using a Mach–Zhender interferometer (Hutchinson 1987), shown schematically in Fig. 5.2. The device is a two-beam system with the plasma under study arranged to be in only one of the arms. A phase change,

$$\Delta\phi = \int (\mu - 1)\,\frac{\omega}{c}\,\mathrm{d}l, \tag{5.6}$$

between the arms is related to the refractive index of the plasma. Here c is the velocity of light and the integral is limited to that part of the interferometer lying in the plasma. Measurement of the phase shift $\Delta\phi$ can thus be used to provide an estimate of the mean refractive index of the plasma along the line of sight. Equation (5.4) can be rewritten in the form

$$\mu^2 = 1 - \frac{n_\mathrm{e}}{n_\mathrm{c}}, \tag{5.7}$$

where the critical wave cut-off density is

$$n_\mathrm{c} = \frac{\omega^2 m_\mathrm{e} \epsilon_0}{e^2}. \tag{5.8}$$

By selecting a sufficiently high frequency ω for the probing wave, one can arrange that $n_\mathrm{e} \ll n_\mathrm{c}$ allowing (5.7) to be expanded to give $\mu \approx 1 - \frac{1}{2} n_\mathrm{e}/n_\mathrm{c}$.

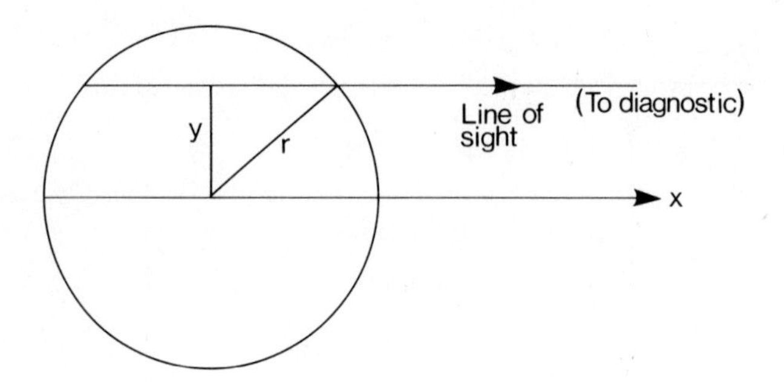

Fig. 5.3. Geometry assumed for the Abel inversion problem. The diagnostic could be used to measure, for example, the number density of electrons along the line of sight passing through a tokamak plasma.

The interferometer signal gives us a measure of the simple chord-averaged electron density,

$$\Delta\phi = \frac{\omega}{2cn_\mathrm{c}} \int n_\mathrm{e} \, \mathrm{d}l. \tag{5.9}$$

A practical version of the Mach–Zhender interferometric technique, based on the principle outlined above, has been described by Magyar (1981).

5.2.2 The discrete Abel inversion

A recurrent problem in plasma diagnosis is to deduce local values of the quantity of interest from the available chordal measurements. Naturally, a solution to this problem is of much wider application than just to plasma diagnostics. We start by considering the geometry shown in Fig. 5.3. Given a set of line integral measurements $\bar{n}_\mathrm{e}(y)$ of a spatial distribution of electrons, we wish to determine the unknown density profile $n_\mathrm{e}(r)$ in the expression

$$\bar{n}_\mathrm{e}(y) = \int n_\mathrm{e}(r) \, \mathrm{d}x, \tag{5.10}$$

using lines of sight at various heights y above the midplane of the plasma.

The geometry shown in Fig. 5.3 could represent, for example, a cross-section through a tokamak plasma. Initially let us assume the measured data to be ideal and noise-free. With lines of sight of the type shown, we construct a set of discrete radial shells and we assume the unknown number density of electrons n_{ej} to be constant in the jth radial shell. The discrete form of (5.10) is

$$\bar{n}_{ek} = \sum_{j=0}^{j_m} L_{kj} n_{ej}, \tag{5.11}$$

where L_{kj} is the incremental length of the kth line of sight intersecting the jth shell and j_m is the maximum shell index number corresponding to the total number of independent line-integral measurements N. When $j_m = N$, the unknown set of discretized electron densities $\{n_{ej}\}$ can be obtained from the matrix equation

$$\{\bar{n}_{ek}\} = \mathbf{L} \cdot \{n_{ej}\} \tag{5.12}$$

by direct inversion:

$$\{n_{ej}\} = \mathbf{L}^{-1} \cdot \{\bar{n}_{ek}\}, \tag{5.13}$$

where

$$\mathbf{L} = \begin{pmatrix} L_{11} & L_{12} & \dots & L_{1N} \\ L_{21} & L_{22} & \dots & L_{2N} \\ \vdots & \vdots & L_{kj} & \vdots \\ L_{j_m 1} & L_{j_m 2} & \dots & L_{j_m N} \end{pmatrix}. \tag{5.14}$$

With matrix inversion, solution of the linear Abel problem is robust when the measured data are noise-free and gives an effective radial resolution on the reconstructed density profile equivalent to the line of sight chord spacing. However, when only a sparse set of measurement chords is available, the quality of the reconstruction is limited, particularly when the data is corrupted by noise. The form of (5.13) and (5.14) shows how errors in the data will propagate from the outermost to the innermost region of the reconstruction. This effect causes the most interesting (central) regions of the profile to be the most severely in error. Linear inversion also gives no means of suppressing negative-valued solutions which, in the present example of counting electrons, would clearly be unphysical. Indeed, as is shown in the example inversions of both trial and real tokamak data below, the linear method can produce negative-valued solutions with realistic data.

5.2.3 The maximum entropy method

We now examine the non-linear maximum entropy (MaxEnt) method of analysis and see how it can give an improved solution. The method embodies a radically different approach to solving our Abel inversion problem. Instead of starting with the measured data and back-transforming it linearly, we start, instead, by considering the set of possible electron density profiles which, when transformed forward (using $\mathbf{L}$), agree with the measured data. From this set (whose Abel transforms agree with the data and its errors), we select a representative member for display. In the MaxEnt formulation, the representative member is the (positive-valued) electron density profile which has the maximum configurational entropy

$$S = -\sum_j p_j \ln \frac{p_j}{m_j}, \tag{5.15}$$

where

$$p_j = \frac{n_{ej}}{\sum_j n_{ej}} \tag{5.16}$$

and m_j is an initial estimate or default level. The configurational entropy S is a measure of the missing information in a particular arrangement $\{n_{ej}\}$ of electron density amongst the cells j. By maximizing S subject to the constraint that the Abel transform of the electron distribution fits the measured data, the resulting reconstruction will contain the least amount of configurational structure (in the sense defined in (5.15)) and yet will remain consistent with the measured data and its associated errors. We note also that the logarithmic form appearing in (5.15) automatically ensures that the reconstruction will be positive-valued. The displayed MaxEnt reconstruction will therefore contain the least amount of spurious structure, artefacts and noise and only such information as is necessary to fit the measured data. Consistency of the MaxEnt solution with the data can be obtained by using the χ^2 statistic,

$$\chi^2 = \sum_k \frac{1}{\sigma_k^2}\left(\sum_j L_{kj} n_{ej}^{\mathrm{m}} - \bar{n}_{ek}\right)^2 \tag{5.17}$$

where $\{n_{ej}^{\mathrm{m}}\}$ is a set of 'mock' data used during iteration and σ_k^2 the variance (due to noise) on the kth datum. Starting with a constant initial default level, an iterative MaxEnt algorithm (Skilling 1984) was used to find the constrained maximum of S. Iteration was stopped when a value of χ^2/N of 1 was reached.

5.2.4 Results of Abel inversion

To test MaxEnt and compare results with linear inversion, both methods have been used to reconstruct four simulated and representative radial electron density distributions (Fig. 5.4). The advantage of using simulated data is that it becomes possible to judge the fidelity of the MaxEnt reconstruction by direct comparison with the original object. A total of $N = 24$ parallel lines of sight was used to generate line-integral data for each density profile. Then a constant level of normally distributed random noise was added to each simulated line-integral value to simulate random

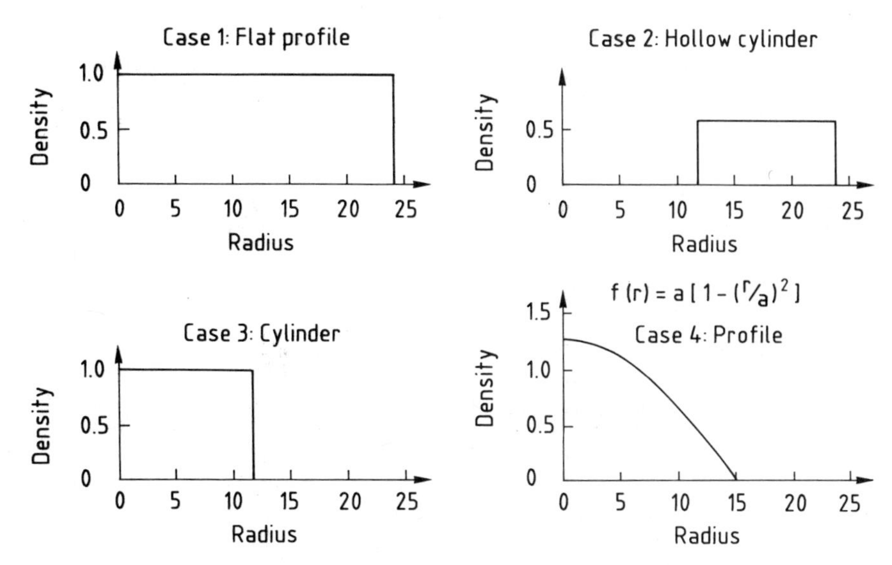

Fig. 5.4. The four trial functions of electron density versus radius used to generate the simulated line-integral data in the Abel inversion problem.

errors on real experimental data. Results are shown in Fig. 5.5. In the flat-profile example, both methods gave reasonable reconstructions; however, the MaxEnt solution also showed noise suppression. The hollow cylinder example could simulate radiation, in a single atomic line, arising from an impurity species in one radial band of the tokamak. Here, again, MaxEnt showed a superior noise performance, particularly in the zero signal region $0 < r < 12.0$, where the matrix solution displayed large amplitude positive and negative fluctuations. Similar features were also found for the case of the cylinder and the plasma profile. By comparing both MaxEnt and matrix solutions with the original electron density distributions of Fig. 5.4, it is possible to derive an effective signal-to-noise ratio on the reconstructions (Table 5.1). In each case the signal-to-noise ratio on the linear matrix reconstruction was less than the maximum on the data, illustrating the expected amplification of noise described above. However, the signal-to-noise ratio in the MaxEnt reconstructions was larger, on average, by about 70% than that obtained with matrix inversion.

The method has also been applied to Abel inversion of real visible light intensity data measured in the DITE (Paul *et al.* 1981) tokamak during experiments to test a magnetic bundle divertor, a device designed to reduce the impurity content of the plasma by diverting part of the plasma flow to an external dump plate. Line-of-sight spectroscopic data were obtained using a monochromator (tuned to CIII impurity line radiation) and

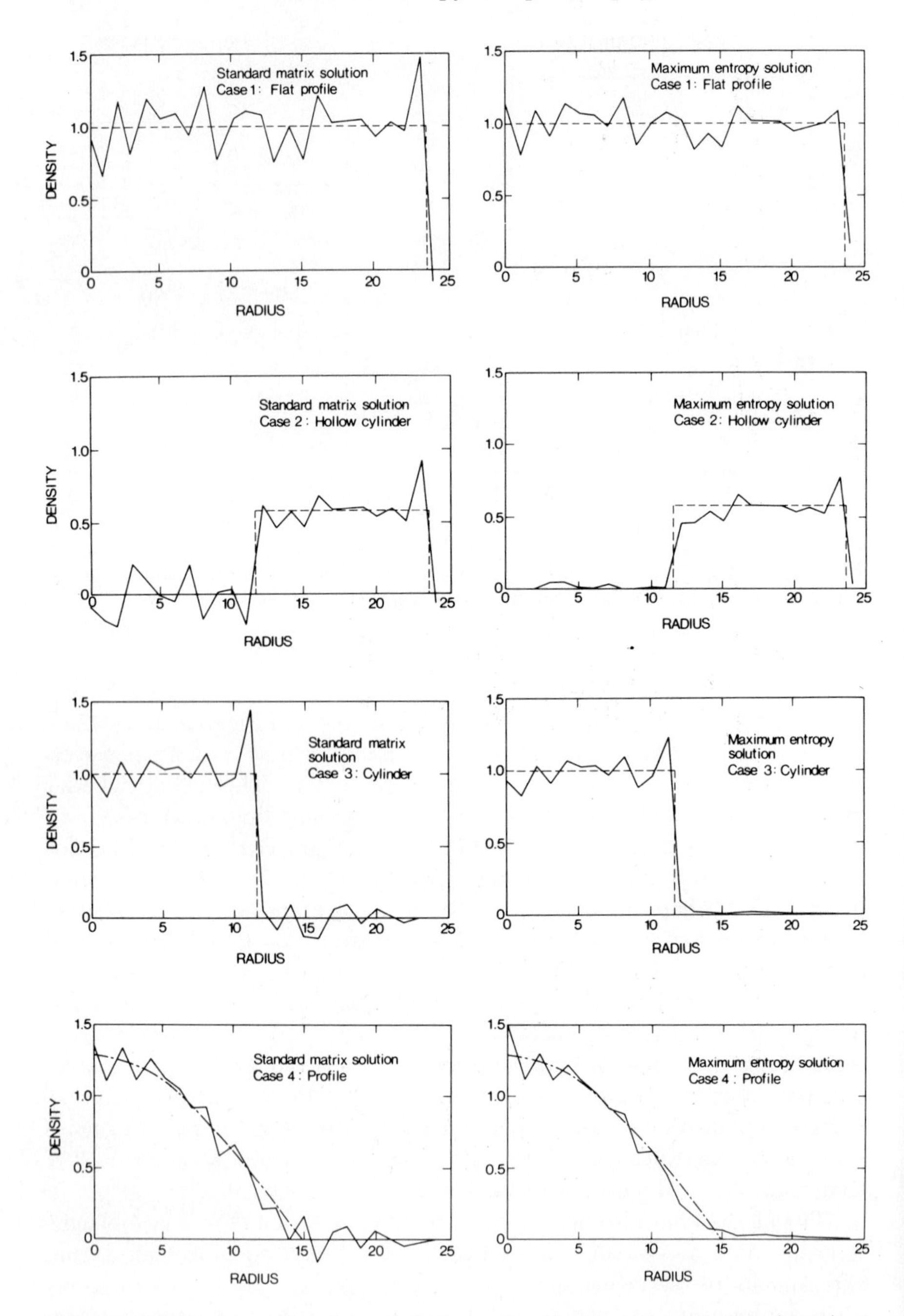

Fig. 5.5. Reconstructions by matrix inversion (left) and by MaxEnt (right) of the four test cases of Fig. 5.4. The peak signal-to-noise ratio on the simulated line-integral data was 20 : 1 in each case.

Table 5.1. Signal-to-noise ratios on the final Abel inversions for a peak signal-to-noise ratio of 20 : 1 on the simulated line-integral data.

Case	Matrix inversion	MaxEnt
1. Flat profile	5.5	9.6
2. Hollow cylinder	4.4	9.2
3. Cylinder	8.5	14.0
4. Plasma profile	15.0	17.0

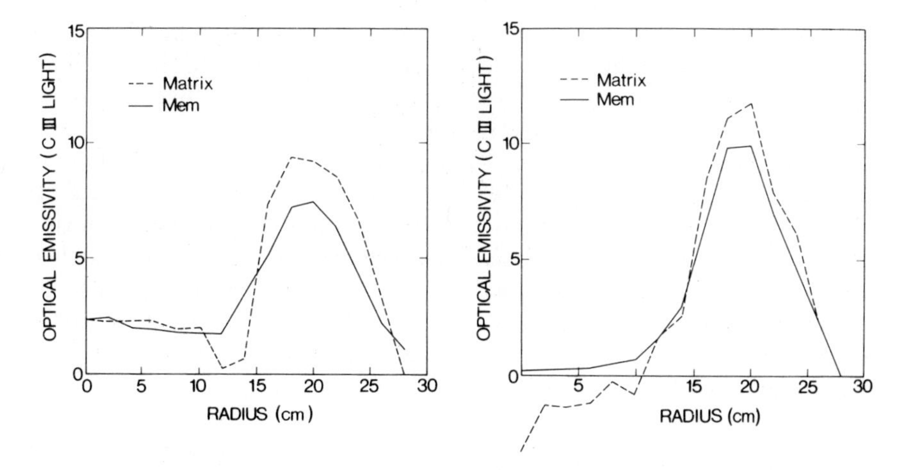

Fig. 5.6. Abel inverted profiles using both matrix inversion and MaxEnt reconstruction of visible CIII radiation emission in the DITE tokamak with the bundle divertor switched off (limiter configuration) (left) and switched on (right).

the plasma was spatially scanned using a rotating mirror device. Comparisons of impurity radiation behaviour were made with the magnetic divertor switched either on or off. The results of Abel inversion of the data are shown in Fig. 5.6. In the limiter (divertor-off) configuration (left) both the matrix and MaxEnt inversions showed carbon light emission to peak near the plasma boundary (radius $r = 20\,\text{cm}$). The fluctuation in emission at $r = 12\,\text{cm}$ seen in the matrix inversion is not seen in the MaxEnt solution and appears to be a result of noise amplification; there was no evidence for a fluctuation in the raw data at the corresponding radius and so it should (and does) not appear in the MaxEnt solution. In the divertor-on case (Fig. 5.6 right), both methods show that the carbon emissivity still peaks at a radius $r \approx 20\,\text{cm}$; however, the matrix solution becomes significantly negative for $r < 8\,\text{cm}$. This is a result of error and noise propagation towards the centre of the reconstruction. The positivity constraint of the MaxEnt method is strikingly clear in this example. By using MaxEnt, the beneficial effect of the divertor in reducing the level of carbon impurities in the plasma centre could now be seen clearly.

In the limit of a very small number of chordal measurements, the degree of coupling between the data and the reconstruction can be weak. Large regions of the profile may become decoupled from the data and the few lines of sight may become strongly weighted to the tangency radii (in the geometry of Fig. 5.3). Simple application of the MaxEnt method with a constant initial default level in this case can lead to the presence of unphysical distortions (at radii corresponding to the tangency radii of the measurements). The problem arising from data coupling may, however, be overcome by including **prior** information on the nature of the distribution of electron density through the default level m_j. Here one can use a plausible model for the prior distribution and encode previous knowledge about the behaviour of electron density distributions in tokamaks before any MaxEnt iteration takes place. It is important to note that MaxEnt will not constrain the reconstruction to be equal to the prior, except in the trivial case where there is a complete absence of data. After iteration, this method does result in the production of a physically plausible profile. A description of the use of a prior with extremely sparse experimental data has been given by Cottrell *et al.* (1985) and a discussion on the use of priors in MaxEnt analysis by Gull (1989).

5.3 Spectroscopic deconvolution

5.3.1 Zeeman splitting of emission lines in a tokamak

Plasma spectroscopy offers another important tool for obtaining data on the state of a plasma. Observation and identification of emission lines from the various atomic constituents in the plasma can yield information on impurity concentrations as well as ion temperatures (from Doppler broadening measurements). More information still comes from observations of the Zeeman splitting of spectral emission lines from atoms in the magnetic field of a tokamak. In a field of strength B, the energy levels of an atom suffer the energy splitting,

$$\Delta E = \Delta(Mg)\mu_{\mathrm{B}}B, \tag{5.18}$$

where M is the total quantum number, g the Landé factor and μ_{B} the Bohr magneton. For spectroscopic observations perpendicular to the magnetic field direction, only the linearly polarized π component (centred on the unshifted line frequency and corresponding to transitions with zero change in the magnetic quantum number, $\Delta m = 0$) is observed. Parallel to the magnetic field, however, the spectral line is split into two circularly polarized σ components which appear symmetrically on both sides of the unshifted line frequency and correspond to $\Delta m = \pm 1$ transitions. In general, the relative intensities of the π and σ components vary according to the viewing direction and the direction of the magnetic field. There are two main applications of this effect in tokamak plasma physics research. Firstly, if the location of the atomic species in the plasma is known (in practice this could be achieved by seeding the plasma with tracer atoms) then the measured Zeeman splitting may be used to determine the local magnetic field strength using (5.18). Conversely, when B is known (typically B can be estimated to within a few percent in a tokamak), then we can use the measured Zeeman splitting to determine the location of the atomic species. This is possible because of the monotonic variation of the toroidal field with major radius (5.3).

5.3.2 Deconvolution of instrumental blurring with MaxEnt

In the experimental example of Zeeman splitting measurements, an optical multichannel analyser was used to survey visible emission from the JET tokamak plasma (Carolan *et al.* 1985). A raw experimental spectrum (together with the measured instrument function) is shown in Fig. 5.7, and reveals a CrI line multiplet and a CII line. There was almost 100% Zeeman modulation of the CrI 4289 Å line but the CrI 4254 Å line (which should also have been split) was apparently featureless. With an expected tokamak field strength $B \approx 3$ tesla, the Zeeman σ components should have been split by $\Delta\lambda \approx 0.75$ Å, a value close to the instrumental resolution (≈ 1 Å). It thus became apparent that the fine spectral structure was being smoothed by the instrument function of the spectrometer. Thus in order to resolve the fine structure, deconvolution was essential.

In a spectrometer, the incoming (original) spectrum f_j is convolved with the instrumental blurring function b_j resulting in the blurred spectrum F_j, that is,

$$F_j = \left[f * b\right]_j \equiv \sum_i b_i f_{j-i}. \tag{5.19}$$

Deconvolution (removal of the effects of this smearing) involves the inverse problem of estimating f_j given both the blurred data F_j and the instrument function b_j. A standard solution to the problem of inverting (5.19) is based on noting that, under Fourier transformation, convolution can be expressed as the product of two Fourier transforms:

$$\text{F.T.}\{F_j\} = \text{F.T.}\{f_j\} \times \text{F.T.}\{b_j\}, \tag{5.20}$$

which shows a possible method of solution. There are two well-known problems with this approach. Firstly, owing to data truncation, artefacts

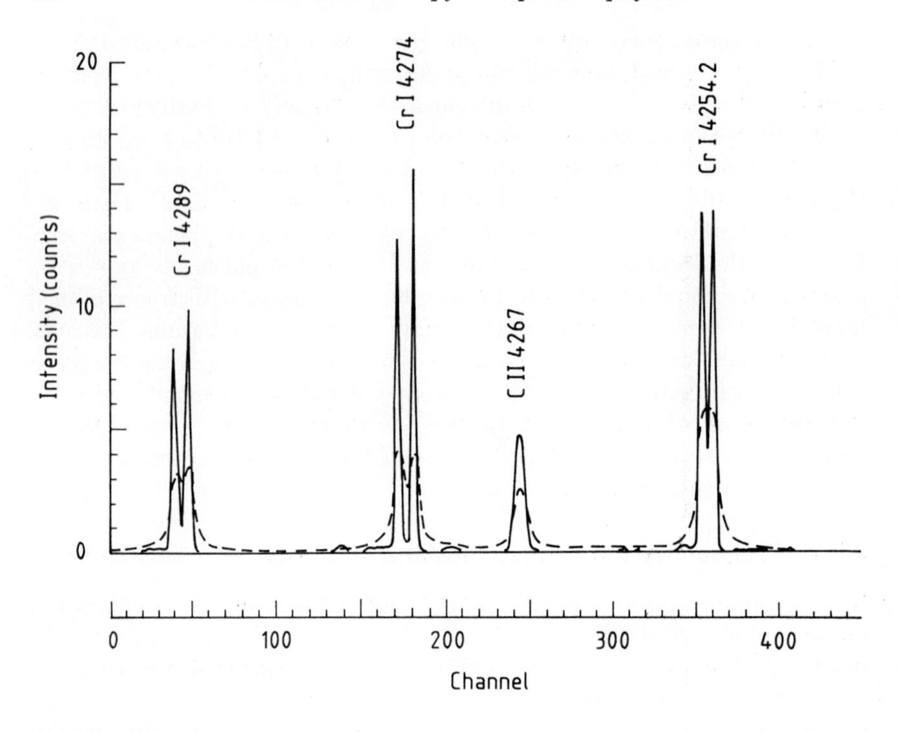

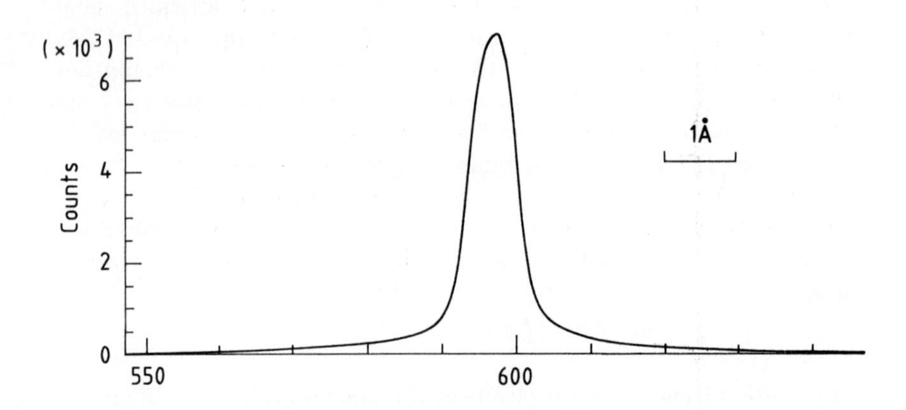

Fig. 5.7. The visible spectrum from a JET tokamak discharge. The raw data (broken line) show some evidence for Zeeman splitting of the chromium and carbon spectral emission lines. The spectrum deconvolved using MaxEnt (solid line) with the spectrometer instrument function (below) reveals that the line CrI 4254 Å is split.

(or 'sidelobes') are introduced. Secondly, the linearly deconvolved spectrum can become negative. In the MaxEnt solution (Fig. 5.7), both of these problems were avoided. MaxEnt was applied to this problem in a way similar to that described in Section 5.2 except that in the forward-transform to the mock spectral data, the convolution operation (5.19) was used. A striking feature of the solution is the discovery that the deconvolved CrI 4254 Å line now shows Zeeman splitting. Moreover, the magnitude of the splitting of the three CrI lines is consistent with a single magnetic field of strength $B = 3.0 \pm 0.1$ tesla (corresponding to the field at radius $R = 2.12\,\mathrm{m}$), and almost exactly coincident with the tangential radius of the viewing chord used in the experiment. Hence, the magnetic field inferred was nearly parallel to the viewing direction, which explained the almost complete absence of the central π component. This can be seen particularly clearly with the deconvolved CrI 4274 Å line.

5.4 Fourier transform spectroscopy

5.4.1 Measurement of electron cyclotron emission from a tokamak plasma

Hot thermal electrons (typically having temperatures in the range $T_{\mathrm{e}} \approx$ 1–20 keV) are confined in the tokamak magnetic field (typically of strength $B \approx$ 1–5 tesla) and gyrate around the magnetic field lines with the electron cyclotron frequency,

$$\omega_{\mathrm{ce}} = \frac{eB}{m_{\mathrm{e}}}. \tag{5.21}$$

The gyrating electrons radiate electron cyclotron emission (ECE) at the fundamental frequency ω_{ce} as well as at its harmonics, $2\omega_{\mathrm{ce}}$, $3\omega_{\mathrm{ce}}$, etc.. Because the magnetic field in a tokamak decreases monotonically (5.3) with the major radius, ω_{ce} also varies with radius. Thus an observation of the frequency of ECE can be used to determine the radial location of a particular group of radiating electrons. Furthermore, as the plasma is optically thick to radiation at the fundamental and first harmonic ($\omega = 2\omega_{\mathrm{ce}}$) of the electron cyclotron frequency, the electrons radiate energy at a level close to that of a black body in this frequency range:

$$I(\omega) \equiv I_{\mathrm{b}}(\omega) = \frac{\omega^2 k T_{\mathrm{e}}}{8\pi^3 c^2}. \tag{5.22}$$

Thus an absolute measurement of the ECE spectrum $I(\omega)$ is equivalent to a measurement of the radial distribution of the electron temperature $T_{\mathrm{e}}(R)$ in the plasma. The physics of electron cyclotron emission has been discussed in more detail by Costley (1982).

Measurements of ECE in tokamaks are often based on the technique of Fourier transform spectroscopy (see, for example, Lipson and Lipson (1981)). For typical tokamak parameters, ECE is observed in the frequency range 60–600 GHz. Experimentally, a two-beam Michelson interferometer is used to measure the autocorrelation function with a typical time sampling rate $\Delta t \approx 10\,\mathrm{ms}$ and frequency resolution $\Delta f \approx 10\,\mathrm{GHz}$. It is, however, well known that the exact spectrum of the incident radiation source can only be determined when the (noise-free) autocorrelation function has been measured for all values of the path difference between the two interfering wave trains. In any practical system, not only do we have to contend with noise in the data but also we have no knowledge of the autocorrelation function beyond a certain maximum path difference $N\delta$ limited by the scanning range of the interferometer.

5.4.2 The autocorrelation function

In general, we require an estimate of the spectrum $I(\omega)$ of a radiation source. This is related to the measured autocorrelation function $A(\tau)$ by the Fourier integral

$$A(\tau) = A_0 + \int_0^\infty I(\omega)\cos(\omega\tau)\,\mathrm{d}\omega, \tag{5.23}$$

where τ is the time delay between the two interfering wave trains and A_0 is a constant background level on the whole interferogram. In practice, the detector output of an interferometer is sampled discretely, giving an estimate of the source spectrum

$$I(\omega) = \frac{2}{\pi}\sum_{n=0}^{N}\left[A(n\Delta\tau - \tau_0) - A_0\right]\cos(\omega n\Delta\tau), \tag{5.24}$$

where τ_0 is an (initially) unknown offset representing the zero path difference (ZPD) correction between the two arms of the interferometer. Fig. 5.8 shows some typical autocorrelation data obtained in experiments (Cottrell *et al.* 1982) on the DITE tokamak using a 16 millimetre-wavelength Michelson interferometer. The autocorrelation function (5.23) is sampled with the discrete time delays $-\tau_0$, $\Delta\tau - \tau_0$, $2\Delta\tau - \tau_0$, ..., $N\Delta\tau - \tau_0$. The maximum spectral resolution is inversely related to the maximum time delay:

$$\Delta\omega = \frac{1}{N\Delta\tau - \tau_0}. \tag{5.25}$$

A spectrum obtained by direct Fourier transformation of the raw data of Fig. 5.8 is shown in Fig. 5.9. Because no weighting was applied prior

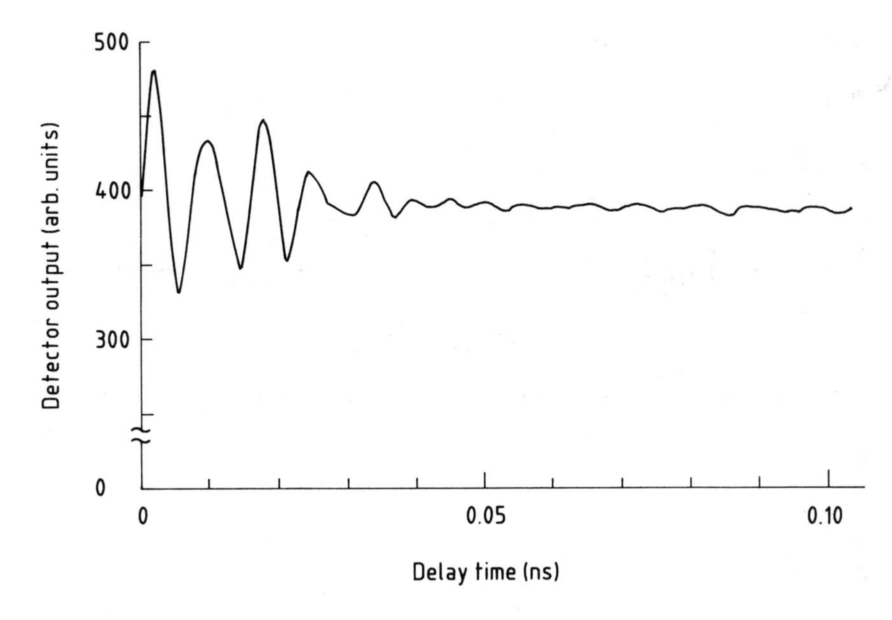

Fig. 5.8. Experimental autocorrelation data for the Fourier transform spectroscopy technique in which electron cyclotron emission spectra from a tokamak plasma were measured. There were $N = 194$ data points in this experiment on a DITE tokamak discharge with a field of 2.2 tesla. The electron cyclotron frequency ω_{ce} was 61 GHz.

to transformation, the frequency resolution obtained is the maximum possible defined in (5.25). However, the spectrum also exhibits unphysical negative regions which are associated with artefacts ('sidelobes') caused by the truncation of the autocorrelation data. To reduce the magnitude of the sidelobes, it is common to weight (or **apodize**) the autocorrelation data prior to transformation. For example, the autocorrelation data $A(\tau)$ of (5.23) can be weighted so that $A^{\text{weighted}}(\tau) = A(\tau)\cos^2(\pi\tau/2N\Delta\tau)$ which tends to zero as $\tau \to N\Delta\tau$, thus removing the discontinuity at the truncation point. However, because short wavelength Fourier components are weighted down strongly in this process, apodization also degrades the spectral resolution (typically by up to a factor of two). So, in choosing an apodization function, one is forced to compromise between resolution and a tolerable level of spurious structure. A standard apodization function is the cosine-squared weighting; the result of applying this function to the data prior to transformation is shown in Fig. 5.10. Although negative sidelobe structure was reduced, resolution was also degraded such that structure on

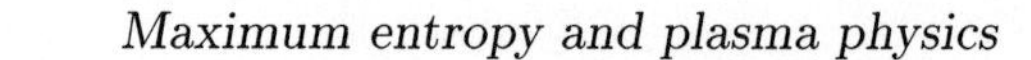

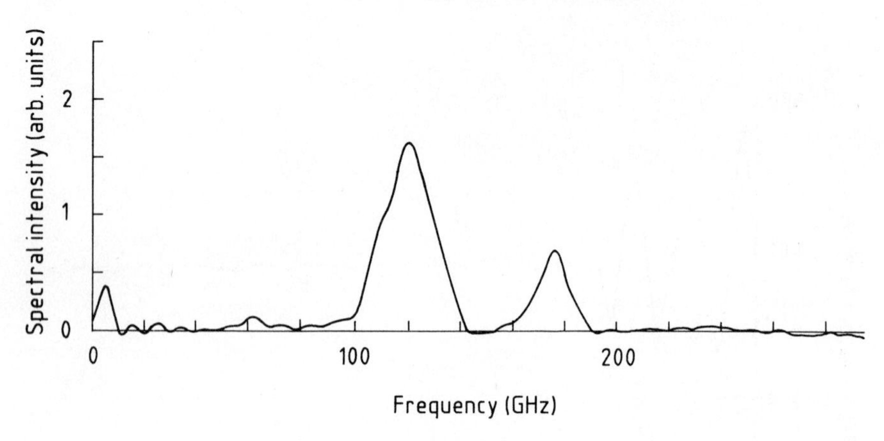

Fig. 5.9. The Fourier transformed spectrum of the unweighted data in Fig. 5.8.

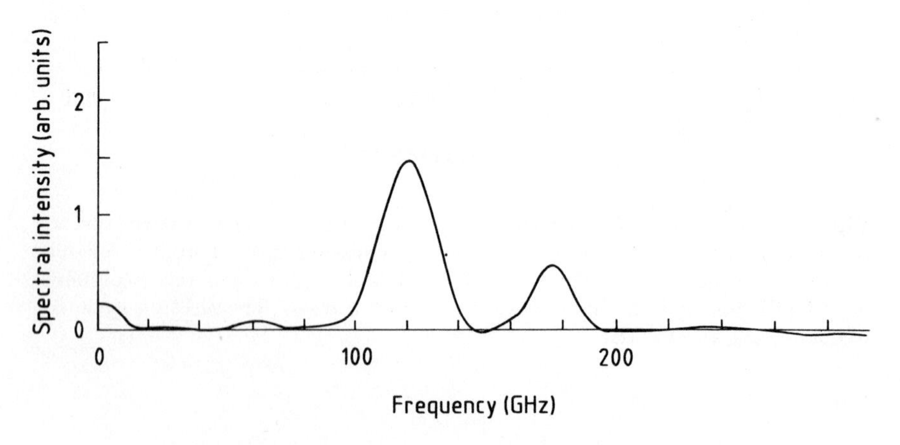

Fig. 5.10. Cosine-squared apodization was applied to the data of Fig. 5.8 prior to Fourier transformation.

the low frequency side of the main peak has been lost and the other peaks broadened.

5.4.3 Autocalibration with MaxEnt

To solve the Fourier transform spectroscopy problem using MaxEnt, we have adopted the same basic method as was used above in the Abel inversion problem. Here the mock autocorrelation data were calculated using the fast Fourier transform and these were subsequently interpolated on to a grid at the exact time delays at which the real data were measured. Iteration was started from a flat spectrum (m_j = constant), using an

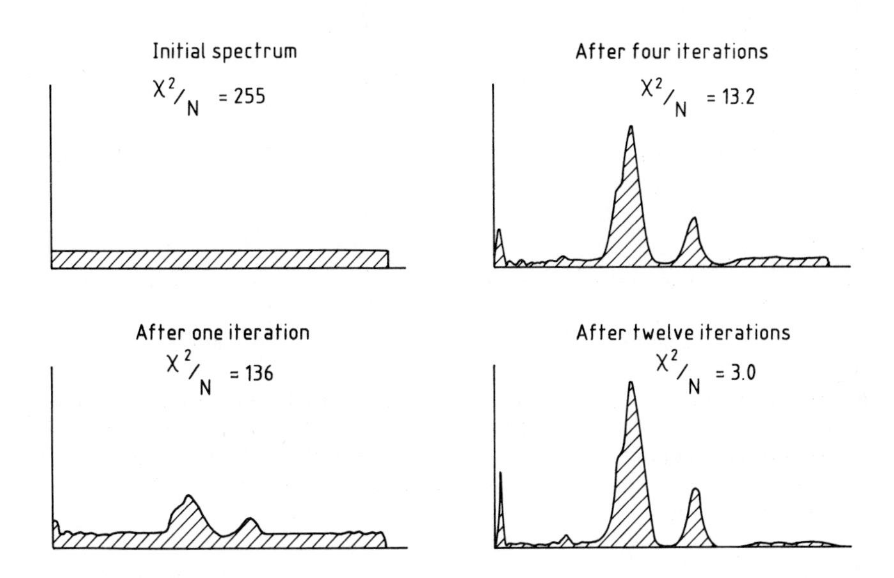

Fig. 5.11. Development of the reconstructed ECE spectrum during MaxEnt iteration on the data of Fig. 5.8 starting from a flat (m_j = constant) spectrum. The oversampling rate was $\alpha = 4$, and the zero path difference correction (ZPD) τ_0 was 4.5 grid points.

initial τ_0 of 4.5 grid points. Fig. 5.11 shows the development of the solution at various stages of iteration. After 12 iterations, $\chi^2/N = 3.0$, but even after 20 iterations it could not be reduced below 2.7. However, the value of χ^2/N, at that stage, was not fully minimized with respect to changes in τ_0, i.e., variations in τ_0 produce variations in the mock data after MaxEnt iteration. Based on the spectra obtained after 15 iterations, χ^2/N was then minimized with respect to τ_0, resulting in an improved value $\tau_0 = 4.6$ grid points, corresponding to the minimum in χ^2/N in Fig. 5.12*a*. One further optimization enabled us to obtain $\chi^2/N = 1.0$ with an optimum ZPD of $\tau_0 = 4.68$ grid points after 16 iterations (Fig. 5.12*b*). In principle this calibration procedure can be applied iteratively. Unless the initial estimate of τ_0 was far from optimum, one or two recalibrations were generally found to be sufficient.

The resulting optimized MaxEnt spectrum is shown in Fig. 5.13. The spectrum has good resolution, shows the (known) structure on the low frequency side of the main harmonic peak and is everywhere positive. In this case, the low frequency structure was believed to be caused by instrumental resonance for which compensation could be applied *a posteriori* by comparison of the plasma spectrum with that of a laboratory black body.

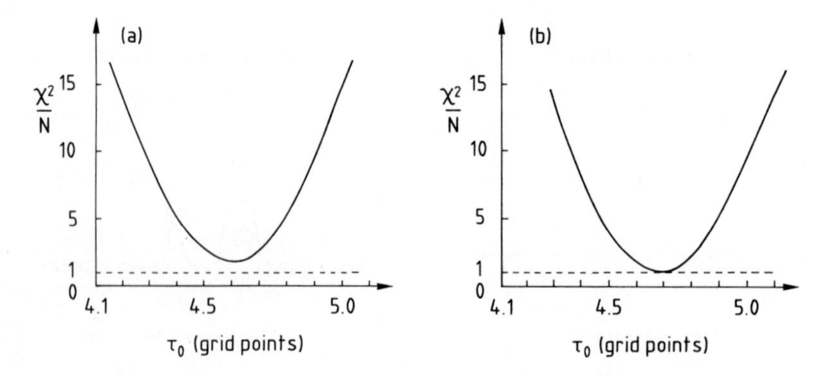

Fig. 5.12. Example of autocalibration with MaxEnt. In this case the (initially) unknown zero path difference (ZPD) correction was determined iteratively. The process involved minimization of χ^2/N during MaxEnt iteration assuming (*a*) the starting ZPD correction $\tau_0 = 4.5$ grid points and (*b*) the final ZPD correction $\tau_0 = 4.68$ grid points.

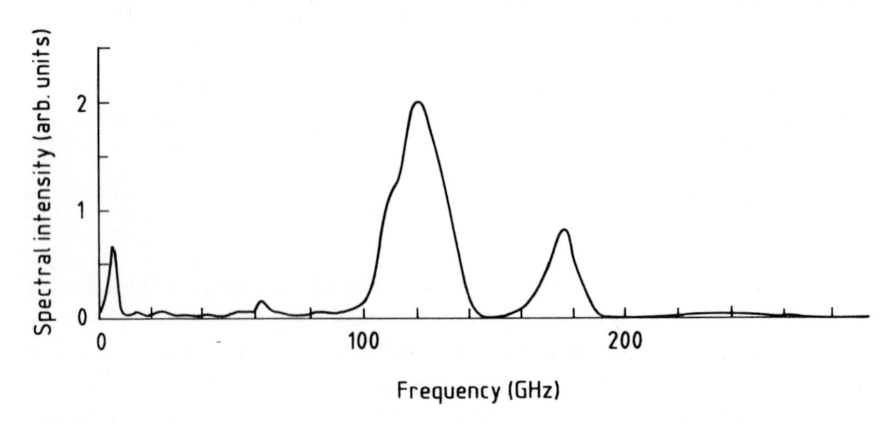

Fig. 5.13. The final, autocalibrated MaxEnt spectrum of electron cyclotron emission.

5.5 Two-dimensional tomography

5.5.1 Neutral beam tomography

During development of the technology of the JET neutral beam injection (NBI) system, it was important to ensure that angular spreading (beam divergence) of the neutral beam particles as they propagate away from the beam source should be below a strict limit, since the size of the beam entrance ports into the tokamak was limited mechanically. At the beam centre (a few metres away from the source) the power density is in the range 10–30 kW cm^{-2}, a level which would melt metallic probes and therefore rule out direct measurements of the beam quality for long-pulse (5–10 second) operation. It therefore became important to develop a different, non-intrusive, method of determining the spatial distribution of the neutral beam in a cross-sectional slice.

The method chosen (Cottrell 1984) was to diagnose the beam optically using observations of the line radiation emitted by fast hydrogen ions and atoms (with energies up to 80 keV per amu and velocities up to $v_b \approx 3 \times 10^6$ m s^{-1}) in two collisional reactions with background molecules in the flight path of the beam:

1. Charge-exchange excitation collisions:

$$\mathbf{H}^+ + \mathrm{H}_2 \rightarrow \mathbf{H}^{0*} \rightarrow \mathrm{H}_\alpha \text{ photon;} \tag{5.26}$$

2. Collisional excitation of ground-state neutral atoms:

$$\mathbf{H}^0 + \mathrm{H}_2 \rightarrow \mathbf{H}^{0*} \rightarrow \mathrm{H}_\alpha \text{ photon,} \tag{5.27}$$

where boldface denotes a fast beam particle and the asterisk an excited neutral (of quantum number $n = 3$ for Balmer alpha emission). By viewing light from the beam atoms at an angle to their direction of motion (typically 10°–40°), emission from fast particles could be separated spectroscopically from background light by means of the finite Doppler shift ($\Delta\lambda \approx 20$ Å) with respect to the H_α rest wavelength ($\lambda_0 = 6563$ Å). The optical emissivity of the fast particles is

$$f(x, y) = n_b(x, y)\, n_0\, \sigma^* v_b, \tag{5.28}$$

where $n_b(x, y)$ and n_0 are, respectively, the number density of fast particles and background hydrogen molecules in the path of the beam and σ^* the Balmer alpha excitation cross-section for reactions (5.26) and (5.27). The x, y coordinates refer to a cross-sectional slice, normal to the direction of propagation. As the beam-plasma is optically thin, a single spectroscopic measurement through the beam is equivalent to a line-integral of the emissivity. Therefore a measurement of $f(x, y)$ can be related to the unknown density of beam atoms $n_b(x, y)$. Spatial scanning of the Balmer alpha beam emission was performed using a number of optical scanners arranged around the periphery of the beam flight tube (Fig. 5.14) with each scanner sampling the H_α light emission using narrow-band interference filters. The measurement geometry therefore corresponds to chord-average samples taken at various angles through the beam.

The unknown two-dimensional neutral beam emissivity function $f(x, y)$ is seen in projection along the viewing lines shown in Fig. 5.14, giving a

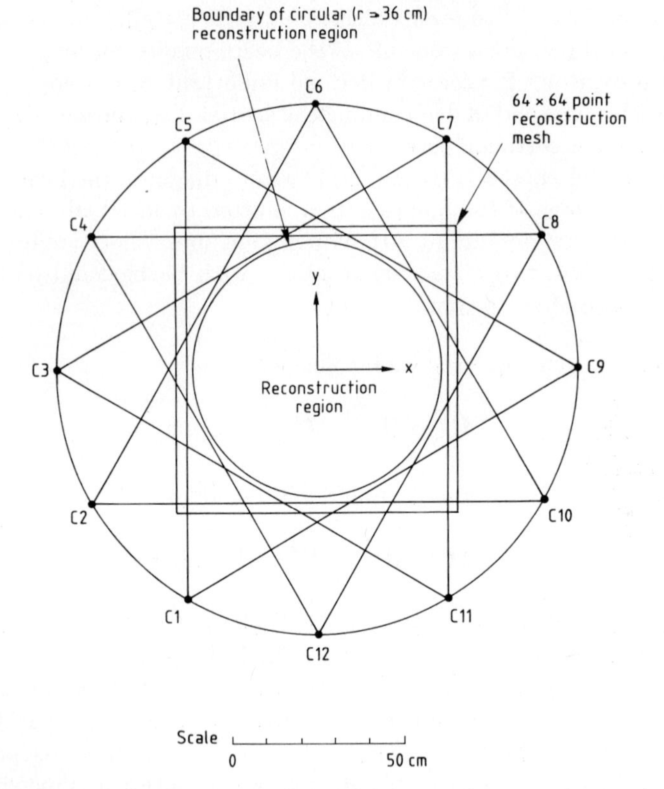

Fig. 5.14. Geometry of neutral beam scanner system projected onto a cross-section through a test beam-line (the beam propagates in the z direction). Camera locations are designated C_1–C_{12} and the lines of sight from each camera are within the viewing limits shown. The inner circle is the reconstruction region lying within the computation mesh. Each camera has an angular resolution of about 1° giving approximately 60 independent lines of sight per camera. Thus there were up to about 720 effective lines of sight available to determine the profile of the neutral beam.

signal response at the kth optical detector ($k = 1, 2, \ldots, M$) of

$$d_k = \iint g_k(\phi) r^{-1} f(x, y)\, \mathrm{d}x\, \mathrm{d}y, \tag{5.29}$$

where $g_k(\phi)$ is the calibrated angular response function of the kth detector and ϕ the angle of the incoming photons with respect to the instantaneous pointing direction of the scanner. The fundamental problem in tomography is an inversion of (5.29) which allows estimation of $f(x, y)$ from the data d_k. Quantizing $f(x, y)$ into a discrete set of values $\{f_j\}$ with $j = 1, 2, \ldots, N$

and calculating matrix element contributions C_{kj} from the jth pixel to the kth detector, we obtain

$$d_k = \sum_j O_{kj} f_j + n_k, \tag{5.30}$$

where n_k is the noise on the kth datum. In the noiseless limit, $n_k = 0$ and (5.30) could, in principle, be inverted linearly.

It was shown by Radon (1917) that $f(x, y)$ cannot be determined completely for a finite number of projections; in this case the reconstruction must be bandwidth-limited. The central slice theorem (Bracewell 1979) can be used to determine the information available when only a finite number of projections is available. The theorem states that, for parallel projections, the Fourier transform of a projection at angle γ is equal to the Fourier transform of the object evaluated along the line at angle γ passing through the origin. Thus to reconstruct an object with diameter D showing spatial structure on the finest scale Δx, we would need

$$n = \frac{\pi D}{\Delta x} \tag{5.31}$$

continuous projections (de Rosier and Klug 1968). For projection data sampled according to the Nyquist criterion, there must be two samples per cycle of the highest spatial frequency present. Thus, of order $2D/\Delta x$ samples per projection are needed.

An important feature of the design of the NBI tomography system was the ability to resolve structure in the peripheral region of the beam with maximum resolution, as it is in this region that the beam focusing quality is most critical. Thus it was essential to use the full angular resolution of the scanners, equivalent to a value $\Delta x = \Delta y \approx 2$ centimetres at the position of the beam. To meet the condition of (5.31) in the geometry shown in Fig. 5.14, we would therefore have needed $n \approx 100$ scanners each operating with full angular resolution. However technical considerations limited us to just 12. Thus in the present application the line-integral data were sparse; we have used MaxEnt to fill in the unmeasured regions of Fourier space.

5.5.2 Solution and autocalibration with MaxEnt

In the MaxEnt inversion, the mock data were generated using the forward transform given by (5.29). Starting from a constant default level, the inversion was calculated on a $64 \times 64 = 4096$ point mesh (Fig. 5.14). During experimental tests it was found that variations in mirror reflectivity, interference filter characteristics and photomultiplier tube efficiencies between the individual scanner units introduced systematic differences in

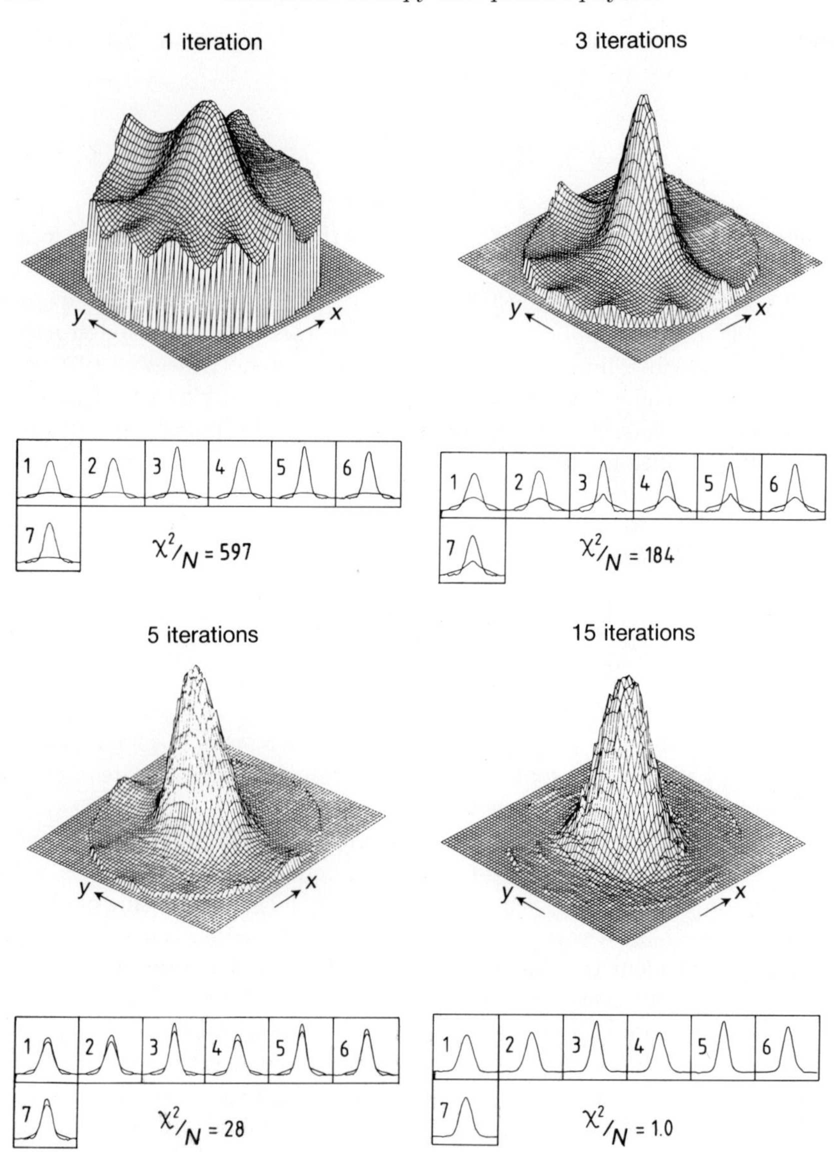

Fig. 5.15. Development of the MaxEnt reconstruction during iteration. Below each isometric projection, the individual camera data are shown (upper curve) as well as the mock data at each iteration (lower curve). After 15 iterations, the two sets of curves merge showing that the data has been fitted by the algorithm, a condition defined by the normalized value of chi-squared reaching unity.

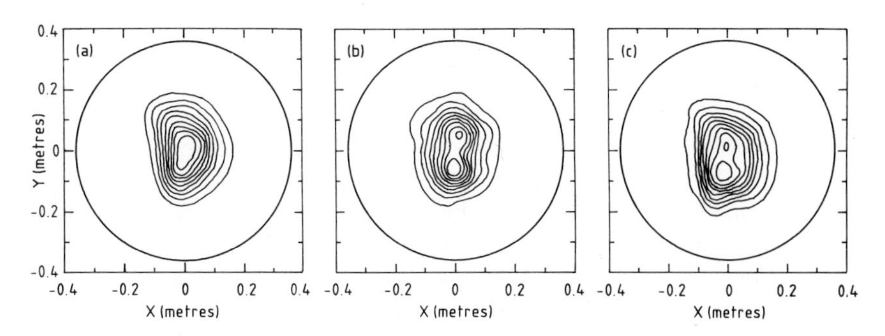

Fig. 5.16. Tomographic reconstructions showing contours of H_α intensity for beam components with energies (a) E_B, (b) $E_B/2$ and (c) $E_B/3$ corresponding to the extraction and collisional break-up of the molecular ions H^+, H_2^+ and H_3^+ from the ion source. The contour levels were normalized to the peak intensity and are at 10%, 20%, ..., 90% of the peak level.

the relative sensitivities. This was corrected in the analysis using the auto-calibration feature of MaxEnt. First, a reconstruction was calculated from an initially calibrated data set. In general, it was found that the value of χ^2/N was, at this stage, not fully optimized with respect to variations of a small set of sensitivity calibration factors, one for each of the cameras. By adjusting these, it was found that it was then possible to reduce the value of χ^2/N, thus improving the consistency of the camera data with a single reconstruction. This is an analogous procedure to that described in Section 5.4.3 for Fourier transform spectroscopy. The development of the MaxEnt solution in the 4096-dimension tomographic problem is shown in Fig. 5.15 after 1, 3, 5 and 15 iterations. Beyond 15 iterations, no significant changes occurred to the solution as χ^2/N had reached unity at the 15th iterate. Worthy of note is the excellent reconstruction obtained with only 7 scanners operational (at positions C_1–C_4, C_6, C_8 and C_{12} on Fig. 5.14) and a signal-to-noise ratio of approximately 100 on the raw camera data. An earlier numerical study (Cottrell 1982) based on simulations of the beam geometry has allowed us to evaluate the fidelity of the results. With only 7 cameras, one would expect some distortion of the contours in Fig. 5.16 at a level about 10% of the peak of the reconstruction. At the 50% level we might also expect to see small distortions related to the presence of noise on the raw data. The double peaks seen in Fig. 5.16*b,c* are, however, real and evidence for them could be seen in the raw data. These peaks are related to the structure of the neutral beam source and to the focusing properties of the electrostatic lenses. Overall, the results indicated that the intrinsic beam divergence was about 0.5° (below the upper limit of the design) and that the spatial spreading of the beam was within the design limits governed by the solid angle of the entrance port into the tokamak chamber.

5.6 Conclusions

The method of maximum entropy has been discussed for four problems in the field of plasma diagnostics. It has been found to offer advantages in the solution of an inverse problem where there is sparse data, the noise on the data cannot be neglected or limitations are imposed by a measuring instrument. Most plasma diagnostic work is beset by a combination of these three problems. Maximum entropy analysis offers a clear framework for dealing with such cases and can help us avoid producing spurious images of positive, additive quantities such as distributions of particle density or numbers of photons per unit wavelength. The study has revealed four clear advantages in using the MaxEnt method over linear inversion methods:

1. There is an enhancement in attainable resolution and signal-to-noise ratio.
2. Inversions are positive and contain no spurious transform-related artefacts. The only configurational structure which appears in an inversion is that for which there is evidence in the original data.
3. The determination of (initially) poorly known instrumental calibration factors is facilitated through the technique of autocalibration.
4. Prior information about the object under reconstruction can be encoded in a simple way through the default level.

There exist other plasma diagnostic problems where the maximum entropy method could be used to advantage. These include reconstruction of other radially varying quantities in the plasma such as neutron and soft X-ray emissivities using imaging cameras and the principles described in Section 5.5. Another possibility is reconstruction of the magnetic equilibrium in a tokamak. The Grad–Shafranov equation (Wesson 1987) is used to determine an optimum choice for the unknown functions of plasma current density $j(r)$ and plasma pressure $p(r)$ by matching predicted and measured magnetic fluxes at the plasma boundary. Often cubic spline functions are used as trial functions in this analysis, which limits the reconstructions to a narrow class for which there may be no evidence in the data. The use of MaxEnt in this context would free the solutions of this limitation and, apart from the benefits of positivity and noise suppression, would enable the use of prior information. As an example, the knowledge that both j and p are zero at the boundary could be encoded simply in terms of the default level.

Acknowledgements

The author would like to thank Dr P. G. Carolan for permission to use the data on Zeeman splitting (adapted here in Fig. 5.7) and Dr S. J. Fielding for permission to use the Abel transform data (Fig. 5.6). It is also a pleasure to thank Drs S. F. Gull and J. Skilling for numerous discussions on MaxEnt theory and applications.

References

Bracewell, R. N. (1979). Image reconstruction in radio astronomy: implementation and applications. In *Image reconstruction from projections,* Topics in Applied Physics, Vol. 32 (ed. G. T. Herman), pp. 81–104. Springer-Verlag, Berlin.

Carolan, P. G., Forrest, M. J., Peacock, N. J., and Trotman, D. L. (1985). Observation of Zeeman splitting of spectral lines from the JET plasma. *Plasma Physics and Controlled Fusion*, **27**, 1101–24.

Costley, A. (1982). Electron cyclotron emission from magnetically confined plasmas. In *Diagnostics for fusion reactor conditions,* Vol. 1, Report 8351-I EN (ed. P. E. Stott, D. K. Akulina, G. G. Leotta, E. Sindoni and C. Wharton), pp. 129–65. Commission of the European Communities.

Cottrell, G. A. (1982). Doppler-shift emission tomography of intense neutral beams. *Journal of Physics E: Scientific Instruments*, **15**, 432–8.

Cottrell, G. A. (1984). Tomography of neutral beams. *Review of Scientific Instruments*, **55**, 1401–9.

Cottrell, G. A., Clarke, W. H. M., and Gull, S. F. (1982). Maximum entropy spectral analysis in Michelson interferometry. In *Diagnostics for fusion reactor conditions,* Report 8351-II EN (ed. P. E. Stott, D. K. Akulina, G. G. Leotta, E. Sindoni and C. Wharton), pp. 177–87. Commission of the European Communities.

Cottrell, G. A., Fairbanks, E. S., and Stockdale, R. E. (1985). Determination of experimental tokamak plasma profiles using maximum entropy analysis. *Review of Scientific Instruments*, **56**, 984–6.

Gull, S. F. (1989). Developments in maximum entropy data analysis. In *Maximum entropy and Bayesian methods, Cambridge, England, 1988* (ed. J. Skilling), pp. 53–71. Kluwer, Dordrecht.

Hutchinson, I. H. (1987). *Principles of plasma diagnostics.* Cambridge University Press.

Lipson, S. G. and Lipson, H. (1981). *Optical physics* (2nd edn). Cambridge University Press.

Magyar, G. (1981). Plasma diagnostics using lasers. In *Plasma physics and nuclear fusion research* (ed. R. D. Gill), pp. 535–50. Academic Press, London.

Paul, J. W. M., Clark, W. H. M., Cordey, J. G., Fielding S. J., Gill, R. D., Hugill, J., *et al.* (1981). The DITE tokamak experiment. *Philosophical Transactions of the Royal Society of London*, **A300**, 535–45.

Pease, R. S. and Bickerton, R. J. (ed.) (1987). The JET project and the prospects for controlled nuclear fusion. *Philosophical Transactions of the Royal Society of London*, **A322**, 1–211.

Radon, J. (1917). Über die Bestimmung von Funktionen durch ihre integralwerte längs gewisser Mannigfaltigkeiten. *Berichte Sächsische Akademie der Wissenschaften. Leipzig, Math.-Phys. Kl.*, **69**, 262–7.

Rosier, D. J. de and Klug, A. (1968). Reconstruction of three dimensional structures from electron micrographs. *Nature*, **217**, 130–4.

Skilling, J. (1984). Cambridge maximum entropy algorithm. In *Maximum entropy and Bayesian methods in applied statistics* (ed. J. H. Justice), pp. 179–93. Cambridge University Press.

Wesson, J. (1987). *Tokamaks,* Oxford Engineering Science Series, Vol. 20. Clarendon Press, Oxford.

6

Macroirreversibility and microreversibility reconciled: the second law

A. J. M. Garrett

Abstract

There is a single consistent resolution of the reversible microdynamics–irreversible macrodynamics problem, stemming from better understanding of the role of probability in physics: the Bayesian viewpoint. The second law of thermodynamics arises on demanding that experiments on thermodynamic variables be reproducible though atomic microvariables differ from run to run. A general formulation for non-equilibrium statistical mechanics is outlined, based on Jaynes' principle of maximum entropy, and the meaning of entropy is clarified using information theory.

6.1 Introduction: the problem

The conundrum facing the pioneers of kinetic theory in the late 19th century was to reconcile the reversible character of Hamiltonian atomic dynamics with the irreversibility of systems of atoms viewed on the large scale. Broken eggs do not spontaneously reform, for example, even though the atoms comprising an egg and its environment behave reversibly. 'Reversible' here means that one cannot tell whether a film of atomic motion is played forwards or backwards: Hamilton's equations

$$\dot{q} = \frac{\partial H(p,q)}{\partial p} \qquad \text{and} \qquad \dot{p} = -\frac{\partial H(p,q)}{\partial q}, \tag{6.1}$$

where $\dot{} \equiv \mathrm{d}/\mathrm{d}t$, q represents the position variables of *all* components of the system and p the canonical momenta conjugate to q, are invariant under the transformation $(q,t) \rightarrow (-q,-t)$. This problem resisted all efforts at solution (Brush 1983) until E. T. Jaynes, building on ideas of Boltzmann and above all Willard Gibbs (1928), solved it in a remarkable series of papers dating from 1957 (Jaynes 1983). We shall state the answer, using as little formal machinery as possible, and then justify it in detail and examine its consequences. Included will be a general formalism for non-equilibrium statistical mechanics.

A typical macroscopic system contains N particles, where $N \approx 10^{23}$, so that in three dimensions the phase space Γ is roughly 6×10^{23}-dimensional. Evolution proceeds by Hamilton's equations, with the system point tracing out a curve in the phase space. Because Hamilton's equations are first order, any point in Γ lies on just one trajectory. In the absence of internal degrees of freedom the trajectory is restricted to a hypersurface of constant energy, constant momentum and constant angular momentum. Knowing seven constants of the motion is not much use, though, when the other $6 \times 10^{23} - 7$—even assuming complete integrability—are unknown.

Define the **microstate** of a system as its position in phase space, specified by $6N$ variables. Define the **macrostate** as something which is specified by only a few **macrovariables**, such as pressure P and temperature T. To every macrostate there therefore correspond many microstates.

Now run a (macro)experiment from initial macrovariable values (P_i, T_i) at time $t_\text{i} = -\tau$ to final values (P_f, T_f) at time $t_\text{f} = 0$. How does this look in phase space? Initially, one out of the set of microstates consistent with the initial macrostate is selected arbitrarily: experiments controlling pressure and temperature do not direct individual atoms. This microstate evolves by Hamilton's equations and at the end is consistent with the final macrostate. Now repeat the experiment, requiring that the result be **macroscopically reproducible**, so that if we start at (P_i, T_i) but with a different atomic disposition (as will almost surely be the case) we still end up with (P_f, T_f). This is a non-trivial requirement additional to the Hamiltonian dynamics. It implies that the time-evolved set of all microstates consistent with the initial macrostate is contained within (or is coincident with) the set of all microstates consistent with the final macrostate: $\Sigma_\text{i}(0) \subset \Sigma_\text{f}(0)$ (see Fig. 6.1). For if not, we could choose an initial microstate corresponding to (P_i, T_i) which evolved to something other than (P_f, T_f).

Macroscopic reproducibility is a *convention* to anyone who knows of atoms; but before the present century, variables such as pressure and temperature could be seen as fundamental, and laws sought among them. Only those variables entering relations reproducible by the technology of the time might be discussed. Once atomism was accepted, reproducibility among the old variables became macroscopic reproducibility, a convention ensuring continuity with the past rather than a dynamical requirement; it is hard for recent generations brought up on atomism to appreciate the idea.

A consequence of Hamilton's equations is that evolution preserves the phase space volume of a connected set of phase points (Liouville's theorem).

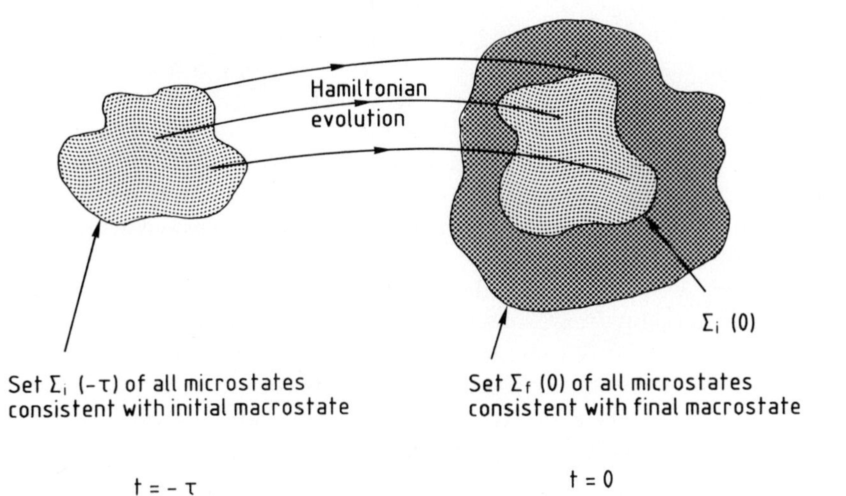

Fig. 6.1. Phase space portrait of a macroreproducible experiment.

Let ρ denote phase space density; then

$$0 = \frac{\mathrm{D}\rho}{\mathrm{D}t} \equiv \frac{\partial \rho}{\partial t} + \dot{p}\frac{\partial \rho}{\partial p} + \dot{q}\frac{\partial \rho}{\partial q} \tag{6.2}$$

$$= \frac{\partial \rho}{\partial t} + \left(-\frac{\partial H}{\partial q}\frac{\partial \rho}{\partial p} + \frac{\partial H}{\partial p}\frac{\partial \rho}{\partial q} \right) \equiv \frac{\partial \rho}{\partial t} + [\rho, H]_{q,p} \tag{6.3}$$

where [,] is the **Poisson bracket**; summation over particles is implied. This is Liouville's equation, and its characteristics are Hamilton's equations. The properties $\rho \geq 0$ and $\int \mathrm{d}\Gamma\, \rho = 1$ propagate unchanged (see, for example, Goldstein (1980)). On denoting by W the phase volume of a set Σ, so that W measures the number of microstates within Σ:

$$W[\Sigma] = \int_\Sigma \mathrm{d}\Gamma, \tag{6.4}$$

we have $W[\Sigma_\mathrm{i}(-\tau)] = W[\Sigma_\mathrm{i}(0)]$, with macroscopic reproducibility implying that $W[\Sigma_\mathrm{i}(0)] \leq W[\Sigma_\mathrm{f}(0)]$. Therefore $W[\Sigma_\mathrm{i}(-\tau)] \leq W[\Sigma_\mathrm{f}(0)]$: the number of microstates consistent with the initial macrostate is less than, or equal to, the number consistent with the final macrostate. Equality implies that the experiment could be reversed, so as always to return to the initial macrostate: $\Sigma_\mathrm{i}(0) = \Sigma_\mathrm{f}(0)$. This thermodynamic or *macro*reversibility is distinct from Hamiltonian microreversibility.

If we take from Boltzmann's tombstone the formula

$$S = k \log W \tag{6.5}$$

as defining an entropy S_Γ, it follows for $k > 0$ and a base of logarithms exceeding one that $S_\Gamma[\Sigma_i(-\tau)] \leq S_\Gamma[\Sigma_f(0)]$: the entropy is a non-decreasing function of time, increasing where there is thermodynamic irreversibility. (As a dimensional convention we take $k = 1$ from now on.) This increase expresses the second law of thermodynamics, with irreversible macroevolution corresponding to continuing loss of *information* over which of the microstates, at a given instant in the past, remain consistent with the incorporated macro-information.

The rest of this chapter is devoted to a fuller derivation, a tour which begins—remarkably—at probability theory. Outstanding problems include: the justification of Boltzmann's epitaph; the location in phase space of boundaries between those microstates which are consistent and inconsistent with a given macrostate; determination of intermediate macrostates and the rate of entropy increase; and reconciliation of (6.5) with the Gibbs entropy

$$S_G \equiv -\int d\Gamma \, \rho \ln \rho, \tag{6.6}$$

which, like any integral $\int d\Gamma \, F(\rho)$, does not change with time in consequence of (6.2). (Given the diverse meanings of 'entropy' catalogued by Jaynes (1980), we point out that anything can be defined provided it is consistent; *usefulness* is the key.)

Before beginning this tour we reiterate what has been achieved. The irreversible increase of entropy has been derived from the reversible Hamiltonian dynamics and the requirement of macroscopic reproducibility, with no question of paradox. This is because the macrovariables—pressure, temperature—are not functions of the microstate, like energy; but are *averages*, weighted by the phase space density ρ, of microstate functions defined on Γ. No amount of philosophy or higher mathematics can rescue us otherwise. The importance of macroreproducibility to the second law has been indicated by Jaynes (1959).

The way in which physics reacted to atomism differs from today's breakdown of determinism in quantum mechanics, in which prediction of observed values proceeds probabilistically. (John Bell (1964) has proved that determinism can be restored only at the expense of locality and causality: see Garrett (1990*a*,*b*).) In quantum theory, the variables retain their interpretations. But, as atomism dawned, the variables themselves—pressure, temperature—frayed, and finally demanded reinterpretation.

Finally, the foregoing analysis makes no reference to the concept of chaos. It asks only that a set of macrovariables be found which exhibit behaviour reproducible to the experimenter's satisfaction. This may be more difficult if the Hamiltonian is chaotic; but chaos, valuable as it is, does not 'explain randomness in statistical mechanics', a phrase which is already troublesome. (What is 'random'?) Nor does it imply that 'even classical systems are non-deterministic', for neighbouring trajectories in phase space also diverge in many non-chaotic systems. Trajectories simply diverge more rapidly in chaotic systems, typically as the exponential of the distance along them.

Let us now return to basics: our study is of systems in which information is discarded so that certainty is unattainable, and the appropriate language for this is probability theory.

6.2 Probability

Probability theory is the system of reasoning applicable in the absence of certainty, and is also known as inductive logic. Today we have a good understanding of how to make inductive inference in a fixed space of outcomes ('possibility space'), but little of how to extend that space, so as to select fresh factors of importance in advance.

There is enduring disagreement over the meaning of probability. Here the **objective Bayesian**—usually just Bayesian—view is presented. (Thomas Bayes was an 18th century Kentish clergyman; Laplace (1820), working soon after, deserves most historical credit.) The Bayesian view is based on demands of consistency; inconsistencies in some alternative viewpoints, notably the frequentist, are discussed in Appendix 6.A.

Probability theory is the way to encode information so as communicate unambiguously in the presence of uncertainty. The quantity called probability is interpretable by such communicants as **degree of consistent belief**, consistency being with the information at hand. For example, it is consistent to assign a low probability to imminent rain on the basis of a fair sky and to assign it a high probability if it is additionally known that a storm is approaching from over the horizon, even though we have not solved the equations of motion for every molecule in the atmosphere. Though it is helpful to regard probability in this way it is not essential, since the calculational exercise proceeds without this interpretation. As soon as belief is mentioned there is a tendency to see the theory as anthropomorphic. But it is immaterial whether probability is assigned by a human brain, an insect's brain or an electronic computer provided each is reasoning consistently in the same space from the same information: the result must be the same. Confusion arises because different brains usually possess different information.

This example also demonstrates that the probability of a proposition is always conditional upon whatever propositions are taken as true. There is no such thing as unconditional probability. (Probability $\frac{1}{2}$ assigned to each face of a coin, prior to uncontrolled tossing, is based on knowing nothing favouring one face over the other.) Of course, if the conditioning propositions assumed true in a probability calculation actually *are* true—

if the sky *is* blue—the assigned probability, is useful in reasoning about the real world. But probability is not a physical attribute of a system, and individuals holding different facts assign different probabilities to the same proposition. This is why it is better to 'assign' than to 'determine' probabilities.

For these reasons, probability assignments are not testable. Should a low probability of rain be assigned from the conditioning information, and should rain subsequently fall, this indicates not the incorrectness of the assignment but the absence of crucial information—that a storm was approaching. Likewise, in uncontrolled coin tossing, one is testing not the proposition that 'the probability of heads is $\frac{1}{2}$', but the proposition that 'the coin was minted evenly'.

We turn now to the quantitative rules governing probability theory, reiterating first the Bayesian view that, to given conditioning information in a given possibility space, there corresponds a unique probability distribution. Rules are needed for *assigning* a distribution in the first place, and for *manipulating* it, so as to update it in the light of fresh information, or to extract the probability of a single proposition from the probability of a joint proposition. Since probability is a mode of logic and not a field-testable theory, consistency is the only criterion available for constructing rules.

We first look at the rules for manipulation, of which there are two: the product rule and the sum rule. With the usual notation that $p(X|Y)$ denotes the probability of X conditioned on Y—these propositions may be composite—and letting $\bar{X}$ denote the negation of X, the product rule is

$$p(X,Y|I) = p(X|Y,I)\,p(Y|I) \tag{6.7}$$

and the sum rule is

$$p(X|I) + p(\bar{X}|I) = 1. \tag{6.8}$$

A pictorial realization of these is given in Fig. 6.2. If the areas of sectors are denoted by a, b, c and d, the product rule (6.7) corresponds to

$$\frac{a}{a+b+c+d} = \frac{a}{a+b} \times \frac{a+b}{a+b+c+d}, \tag{6.9}$$

while the sum rule (6.8) is

$$\frac{a+d}{a+b+c+d} + \frac{b+c}{a+b+c+d} = 1. \tag{6.10}$$

It is stressed that this is an actualization; the rules were first *derived* satisfactorily by R. T. Cox (1946). (Tribus gives the fullest exposition (1969).)

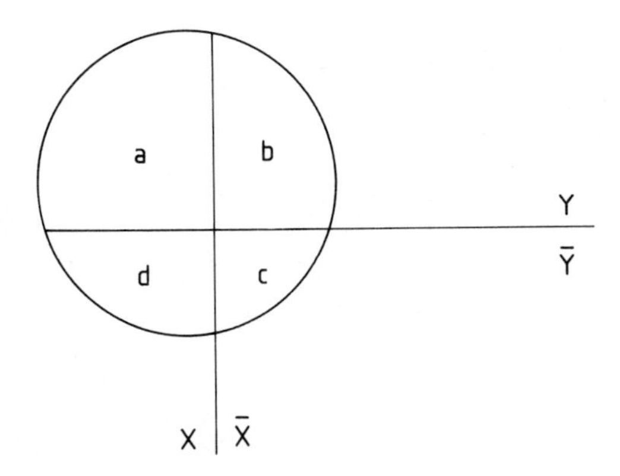

Fig. 6.2. Visualization of the laws of probability.

The product rule emerges in the following way: since $p(X, Y|I) = p(Y, X|I)$ the rule admits two decompositions, so that

$$p(X|Y, I) = p(X|I)\,\frac{p(Y|X, I)}{p(Y|I)}. \tag{6.11}$$

This allows us to update probabilities, here of X in the light of Y, from $p(X|I)$ to $p(X|Y, I)$; in this context $p(Y|X, I)$ is called the **likelihood**. From the Bayesian notion that a probability depends on the totality of its conditioning information, the same results must be attained by updating with *two* fresh pieces of information irrespective of their order; or if the pieces are merged and their joint incorporated in a single update. The updating rule (6.11) satisfies this requirement: update X first with Y as in (6.11), and then the result with Z:

$$p(X|Y, Z, I) = p(X|Y, I)\,\frac{p(Z|X, Y, I)}{p(Z|Y, I)}. \tag{6.12}$$

Now substitute (6.11) into this to give

$$\begin{aligned} p(X|Y, Z, I) &= p(X|I)\,\frac{p(Z|X, Y, I)\,p(Y|X, I)}{p(Z|Y, I)\,p(Y|I)} && (6.13)\\ &= p(X|I)\,\frac{p(Y, Z|X, I)}{p(Y, Z|I)}. && (6.14) \end{aligned}$$

Since the RHS contains Y and Z as their joint, the property is established; only the product rule has been used. To prove the converse, that the product rule (and a family equivalent to it) satisfies the requirement, is more involved (Cox 1946; Tribus 1969).

By combining the sum rule with the product rule, we have

$$p(X,Y|I) + p(X,\bar{Y}|I) = \left[p(Y|X,I) + p(\bar{Y}|X,I)\right] p(X|I) \quad (6.15)$$
$$= p(X|I). \quad (6.16)$$

This is the **marginalizing rule**; it tells us how to extract elementary probabilities from joint probabilities by marginalizing (summing, or averaging) over the unwanted propositions. It comes about from the requirement that unknown details concerning what we don't know should not affect the probabilities of things we do. For suppose that the sum rule adds not to one but to ν, so that the RHS of (6.16) is $\nu p(X|I)$. Now suppose we learn that Y is a composite proposition with components U, V. Then we require that summing over Y give the same result as summing over U, V:

$$p(X,Y|I) + p(X,\bar{Y}|I) = p(X,U,V|I) + p(X,U,\bar{V}|I) + p(X,\bar{U},V|I) + p(X,\bar{U},\bar{V}|I). \quad (6.17)$$

The marginalizing rule reduces the LHS to $\nu p(X|I)$ and the RHS to $\nu^2 p(X|I)$, and equating these gives $\nu = 1$. Again Cox (1946) and Tribus (1969) present formal derivations, including the placement of probabilities in the interval $[0,1]$; probability is a generalization of Boolean, deductive logic from values 0 and 1 to the interval in between. If Boolean algebra is translated into normal algebra it is the same as probability theory in which all propositions have probability 0 or 1. A pleasant aphorism is that the laws derive from requirements that probability depend uniquely on what we do know and not depend on what we don't. Some consequences of Bayesian probability theory are discussed in Appendix 6.B; here, we turn to the problem of probability assignment.

Assignment of probability on a given possibility space also proceeds by consistency. For the sake of simplicity we consider discrete possibility spaces having uniform weighting ('degeneracy') on the possibilities. Generalization is discussed in Appendix 6.B. A crucial concept is the **information content** of a distribution: a sharply peaked distribution for a parameter is clearly more informative than a broad one, with the perfectly sharp limit corresponding to certainty as to the parameter's value. Supposing that we know the expression for the information content of a distribution, probability assignment proceeds by selecting that distribution, out of those consistent with whatever we already know, having the least information content. Consistency with anything known about the distribution ('testable information') must be maintained; while to choose a more informative distribution is to pretend to extra information we do not possess, and to make unwarranted implicit assumptions concerning things we do not know about. Testable information is of the form 'the mean is 5', or should we later wish to estimate the mean, 'the mean is μ'. Samples of a variable are not testable information.

It remains to characterize the information content $\mathcal{I}[p(i)]$ of a distribution $p(i)$ with $i = 1, 2, \ldots, n$. This was done by Claude Shannon (1948), with the famous result

$$\mathcal{I}[p(i)] = \sum_i p(i) \log p(i), \tag{6.18}$$

which varies from $-\log n$ for the uniform distribution $p(i) = 1/n \ \forall i$ to zero for the 'certainty' distribution $p(i) = \delta_{i,i'}$ for some i'. (The base of logarithms is taken to exceed unity.) Shannon's expression comes about by partitioning the possibility space into subspaces, each carrying probability equal to the sum of the probabilities of the elements within; renormalizing the individual distributions in each subspace; and demanding that the information content of the original distribution equal that of the distribution of the subspaces, plus the information content of each subdistribution weighted by the probability of its subspace. Only this recipe is independent of the partitioning, which can be performed in many ways. The criterion leads to a functional equation whose solution is (6.18) (Shannon 1948).

Expression (6.18) has very pleasing properties. Its negative $S[p(i)]$—called **information entropy**—can be proved equal (in base 2 of logarithm) to the expected number of binary (yes/no) questions whose answers take us from our current state of knowledge to one of certainty. The logarithmic form guarantees positivity of the assigned distribution. If there are no constraints apart from normalization the uniform distribution is the result, in line with intuition. As more constraints are brought in, the distribution sharpens.

Distributions are assigned by minimizing $\mathcal{I}[p(i)]$, or by maximizing its negative $S[p(i)]$: the **principle of maximum entropy**. Operationally, this is a routine application of the variational calculus, with constraints incorporated using Lagrange multipliers. Constraints almost always concern expectations of functions defined on the possibility space, when they become linear in the distribution sought and there is a standard formalism for the maximum entropy assignment (Jaynes 1963), given in the next section. The present section, combined with Appendix 6.B, has set forth a self-contained review of the ideas of Bayesian probability theory; Appendix 6.A has indicated the superiority of this over other notions of probability.

6.3 Probability assignment under linear constraints

Suppose that there are m linear constraints in addition to normalization, labelled by j:

$$\langle f_j \rangle \equiv \sum_i p(i) f_j(i) = F_j \qquad \text{for } j = 1, 2, \ldots, m. \tag{6.19}$$

For example, the constraint $4p(1) + 7p(2) + 10p(3) = 8$ corresponds to the constraint function $f(i) = 4\delta_{i,1} + 7\delta_{i,2} + 10\delta_{i,3}$ and $F = 8$. Constraints may be written in vector form $\langle \boldsymbol{f} \rangle = \boldsymbol{F}$. The normalization condition is

$$\langle 1 \rangle \equiv \sum_i p(i) = 1. \tag{6.20}$$

Introduce a Lagrange multiplier λ_0 to handle normalization, and the set $\{\lambda_j\}$ for the other constraints, and make $S[p(i)] - \lambda_0 \langle 1 \rangle - \boldsymbol{\lambda} \cdot \langle \boldsymbol{f} \rangle$ stationary in the usual way. We acknowledge the ubiquity of normalization by renaming its Lagrange multiplier (conventionally) $\ln Z - 1$. Denote the desired distribution $p^{\mathrm{c}}(i)$ and the corresponding multiplier values λ_j^{c}; then, with the conventional base e in (6.18),

$$\begin{aligned} &\delta[S - \lambda_0 \langle 1 \rangle - \boldsymbol{\lambda} \cdot \langle \boldsymbol{f} \rangle] \\ &\quad = \sum_i [-1 - \ln p(i) - (\ln Z - 1) - \boldsymbol{\lambda} \cdot \boldsymbol{f}(i)]\, \delta p(i) \qquad (6.21) \\ &\quad = 0 \ \forall \delta p(i) \qquad \text{at} \qquad p(i) = p^{\mathrm{c}}(i). \qquad (6.22) \end{aligned}$$

The square brackets on the RHS of (6.21) vanish, giving

$$p^{\mathrm{c}}(i) = Z^{\mathrm{c}-1} \exp\bigl(-\boldsymbol{\lambda}^{\mathrm{c}} \cdot \boldsymbol{f}(i)\bigr). \tag{6.23}$$

(This is not written as $p(i|\boldsymbol{F})$ since testable information resides in a different space from conventional conditioning information. Present-day notation is inadequate.) We call this the **canonical distribution**. On substituting it into (6.20), the **partition function** Z is expressed in terms of the Lagrange multipliers:

$$Z^{\mathrm{c}} \equiv Z(\boldsymbol{\lambda}^{\mathrm{c}}) = \sum_i \exp\bigl(-\boldsymbol{\lambda}^{\mathrm{c}} \cdot \boldsymbol{f}(i)\bigr). \tag{6.24}$$

The multipliers are determined by substituting the solution back into the constraints (6.19), giving after manipulation

$$-\frac{\partial \ln Z(\boldsymbol{\lambda}^{\mathrm{c}})}{\partial \boldsymbol{\lambda}^{\mathrm{c}}} = \boldsymbol{F}. \tag{6.25}$$

Equations (6.25) constitute m simultaneous equations for the Lagrange multipliers $\{\lambda_j^c\}$ in terms of the constraints $\{F_j\}$. The canonical distribution may now be envisaged directly in terms of the f_j and their expectation variables F_j. Much of what follows is simply an exercise in implicit algebraic dependence.

If we view the multipliers as independent variables, then

$$\mathrm{d}\big(\ln Z(\boldsymbol{\lambda}^c)\big) = -\boldsymbol{F} \cdot \mathbf{d}\boldsymbol{\lambda}^c \tag{6.26}$$

from (6.25). The stationary value S^c of the information entropy is

$$S^c = \ln Z(\boldsymbol{\lambda}^c) + \boldsymbol{\lambda}^c \cdot \boldsymbol{F}. \tag{6.27}$$

This may be imagined as a function either of the multipliers or of the constraint variables, with (6.27) the Legendre transformation between. From (6.26) and (6.27),

$$\mathrm{d}S^c = \boldsymbol{\lambda}^c \cdot \mathbf{d}\boldsymbol{F} \tag{6.28}$$

whence, treating S^c as a function of the constraints,

$$\frac{\partial S^c[\boldsymbol{F}]}{\partial \boldsymbol{F}} = \boldsymbol{\lambda}^c. \tag{6.29}$$

The vector $\boldsymbol{\lambda}^c$ is perpendicular to surfaces of constant maximized-entropy in constraint space.

Any stationary point discovered is unique and maximal; to prove this, integrate the inequality $y^{-1} \gtrless y^{-2}$ with $0 < y \gtrless 1$ from 1 to x to give $\ln x \geq (1 - x^{-1}) \forall x > 0$, and apply with $x = p(i)/p^c(i)$ to find

$$\sum_i p(i) \ln\left(\frac{p(i)}{p^c(i)}\right) \geq \sum_i p(i)\left(1 - \frac{p^c(i)}{p(i)}\right) = 0. \tag{6.30}$$

Substitution of the canonical distribution (6.23) into the LHS and comparison with (6.27) gives $S[p(i)] \leq S[p^c(i)]$, with equality only at $p(i) = p^c(i)\ \forall i$. Uniqueness is a consequence of the concavity property of information entropy:

$$\eta S\big[p^{(1)}(i)\big] + (1-\eta) S\big[p^{(2)}(i)\big] \leq S\big[\eta p^{(1)}(i) + (1-\eta) p^{(2)}(i)\big], \tag{6.31}$$

where $0 \leq \eta \leq 1$. Uniqueness may fail if constraints are nonlinear in the distribution. Introduction of a fresh constraint reduces the maximized entropy, since the revised canonical distribution remains consistent with the original constraints but no longer corresponds to them alone. The *existence* of a maximum depends on the multipliers being such that $\exp(-\boldsymbol{\lambda}^c \cdot \boldsymbol{f})$ is real and normalizable, and on compatibility of the constraints (for example, $\langle i^2 \rangle \geq \langle i \rangle^2$).

Maximized entropy is concave in constraint space: the distribution $\eta p^{(1)}(i) + (1-\eta)p^{(2)}(i)$ corresponds to constraints $\eta \boldsymbol{F}^{(1)} + (1-\eta)\boldsymbol{F}^{(2)}$, and so by maximum entropy

$$S\left[\eta p^{(1)}(i) + (1-\eta)p^{(2)}(i)\right] \leq S^{\mathrm{c}}\left[\eta \boldsymbol{F}^{(1)} + (1-\eta)\boldsymbol{F}^{(2)}\right]. \tag{6.32}$$

Here it is supposed that $\boldsymbol{F}^{(1)}$ and $\boldsymbol{F}^{(2)}$ correspond to the same constraint functions $\boldsymbol{f}$. In combination with (6.31), this gives

$$\eta S\left[p^{(1)}(i)\right] + (1-\eta)S\left[p^{(2)}(i)\right] \leq S^{\mathrm{c}}\left[\eta \boldsymbol{F}^{(1)} + (1-\eta)\boldsymbol{F}^{(2)}\right], \tag{6.33}$$

with $0 \leq \eta \leq 1$ and the concavity of $S^{\mathrm{c}}[\boldsymbol{F}]$ following upon taking the distributions $p^{(1)}$ and $p^{(2)}$ as canonical.

In order to transform between the multiplier set of variables $\{\lambda_j^{\mathrm{c}}\}$ and the constraint set $\{F_j\}$, we need the matrix elements $A_{jk} \equiv \partial F_j / \partial \lambda_k^{\mathrm{c}}$ and $B_{jk} \equiv \partial \lambda_j^{\mathrm{c}} / \partial F_k$. Matrices A and B are mutually inverse. From (6.26),

$$\begin{aligned}
-A_{jk} &= \frac{\partial^2 \ln Z(\boldsymbol{\lambda}^{\mathrm{c}})}{\partial \lambda_j^{\mathrm{c}} \partial \lambda_k^{\mathrm{c}}} && (6.34)\\
&= \frac{1}{Z^{\mathrm{c}}} \frac{\partial^2 Z^{\mathrm{c}}}{\partial \lambda_j^{\mathrm{c}} \partial \lambda_k^{\mathrm{c}}} - \left(\frac{1}{Z^{\mathrm{c}}} \frac{\partial Z^{\mathrm{c}}}{\partial \lambda_j^{\mathrm{c}}}\right)\left(\frac{1}{Z^{\mathrm{c}}} \frac{\partial Z^{\mathrm{c}}}{\partial \lambda_k^{\mathrm{c}}}\right) && (6.35)\\
&= \sum_i p^{\mathrm{c}}(i) f_j(i) f_k(i) - \sum_i p^{\mathrm{c}}(i) f_j(i) \sum_{i'} p^{\mathrm{c}}(i') f_k(i') && (6.36)\\
&= \langle f_j f_k \rangle_{\mathrm{c}} - \langle f_j \rangle_{\mathrm{c}} \langle f_k \rangle_{\mathrm{c}} && (6.37)\\
&= \langle (f_j - \langle f_j \rangle_{\mathrm{c}})(f_k - \langle f_k \rangle_{\mathrm{c}}) \rangle_{\mathrm{c}}, && (6.38)
\end{aligned}$$

where we write F_j temporarily as $\langle f_j \rangle_{\mathrm{c}}$. The matrix A is called the covariance matrix. It is symmetric and non-positive definite:

$$\begin{aligned}
-\mathbf{d}\boldsymbol{\lambda}^{\mathrm{c}} \cdot \mathbf{d}\boldsymbol{F} &= -\mathbf{d}\boldsymbol{\lambda}^{\mathrm{c}} \cdot \mathsf{A} \cdot \mathbf{d}\boldsymbol{\lambda}^{\mathrm{c}} && (6.39)\\
&= \langle (\boldsymbol{f} \cdot \mathbf{d}\boldsymbol{\lambda}^{\mathrm{c}} - \langle \boldsymbol{f} \cdot \mathbf{d}\boldsymbol{\lambda}^{\mathrm{c}} \rangle_{\mathrm{c}})^2 \rangle_{\mathrm{c}} && (6.40)\\
&\geq 0. && (6.41)
\end{aligned}$$

Inequality is strict unless the constraint functions $f_j(i)$ are linearly dependent. Next, from (6.29),

$$B_{jk} = \frac{\partial^2 S^{\mathrm{c}}[\boldsymbol{F}]}{\partial F_j \partial F_k} \tag{6.42}$$

and

$$\mathbf{d}\boldsymbol{\lambda}^{\mathrm{c}} \cdot \mathbf{d}\boldsymbol{F} = \mathbf{d}\boldsymbol{F} \cdot \mathsf{B} \cdot \mathbf{d}\boldsymbol{F} = \mathbf{d}\boldsymbol{\lambda}^{\mathrm{c}} \cdot \mathsf{A} \cdot \mathbf{d}\boldsymbol{\lambda}^{\mathrm{c}}. \tag{6.43}$$

The matrix $\mathbf{B}$ is also non-positive definite, in line with the concavity of $S^{\mathrm{c}}[\boldsymbol{F}]$.

We are now in a position to make inferences about other quantities which are not constrained. Define a function $\phi(i)$ on the possibility space. Then

$$-\frac{\partial\langle\phi\rangle_{\mathrm{c}}}{\partial\boldsymbol{\lambda}^{\mathrm{c}}} = -\sum_i \frac{\partial p^{\mathrm{c}}(i)}{\partial\boldsymbol{\lambda}^{\mathrm{c}}}\phi(i), \tag{6.44}$$

$$= \big\langle(\boldsymbol{f} - \langle\boldsymbol{f}\rangle_{\mathrm{c}})(\phi - \langle\phi\rangle_{\mathrm{c}})\big\rangle_{\mathrm{c}} \tag{6.45}$$

upon simplification.

Higher moment relations are derived by differentiating $Z(\boldsymbol{\lambda}^{\mathrm{c}})$ further, employing (6.45) with appropriate choice of ϕ to remove derivatives.

Generalizations are now examined. The first, to a higher-dimensional possibility space, is routine. A notable result is that if the constraints are the marginal distributions for each individual element, then the maximum entropy distribution is the product of these. The second generalization is to a continuous possibility space $\{x\}$, upon which there is a measure $m(x)$ induced by symmetry considerations (see Appendix 6.B). We maximize now

$$S = -\int \mathrm{d}x\, P(x)\ln\left(\frac{P(x)}{m(x)}\right) \tag{6.46}$$

to find the probability density $P(x)$. The Gaussian distribution, for example, is the maximum entropy distribution corresponding to uniform measure on $(-\infty, \infty)$ with the first two moments—the mean and the variance—constrained. Another important generalization is to continuous constraint label j. There is then an infinity of constraints, one for each value of j, and our results go over to the functional calculus. For example (6.25) becomes

$$-\frac{\delta \ln Z[\lambda^{\mathrm{c}}(j)]}{\delta\lambda^{\mathrm{c}}(j)} = F(j). \tag{6.47}$$

So far we have considered variations in the constraint variables F_j. We now additionally let the constraint functions $f_j(i)$ vary, by supposing they depend on auxiliary parameters $\alpha_1, \alpha_2, \ldots,$ so far suppressed, which alter. What are the induced changes in $\ln Z^{\mathrm{c}}$ and in S^{c}? Our starting equations are

$$p^{\mathrm{c}}(i,\boldsymbol{\alpha}) = {Z^{\mathrm{c}}}^{-1}\exp\big(-\boldsymbol{\lambda}^{\mathrm{c}}\cdot\boldsymbol{f}(i,\boldsymbol{\alpha})\big), \tag{6.48}$$

$$Z^{\mathrm{c}} \equiv Z(\boldsymbol{\lambda}^{\mathrm{c}},\boldsymbol{\alpha}) = \sum_i \exp\big(-\boldsymbol{\lambda}^{\mathrm{c}}\cdot\boldsymbol{f}(i,\boldsymbol{\alpha})\big), \tag{6.49}$$

$$-\left.\frac{\partial\ln Z^{\mathrm{c}}}{\partial\boldsymbol{\lambda}^{\mathrm{c}}}\right)_{\boldsymbol{\alpha}} = \boldsymbol{F} \tag{6.50}$$

$$\text{and}\quad S^{\mathrm{c}} = \ln Z^{\mathrm{c}} + \boldsymbol{\lambda}^{\mathrm{c}}\cdot\boldsymbol{F}. \tag{6.51}$$

We need the partial derivatives of $\ln Z^{\rm c}$ and $S^{\rm c}$. First, it is readily shown that

$$-\left.\frac{\partial \ln Z^{\rm c}}{\partial \alpha_k}\right)_{\boldsymbol{\lambda}^{\rm c}} = \boldsymbol{\lambda}^{\rm c} \cdot \left\langle \frac{\partial \boldsymbol{f}}{\partial \alpha_k} \right\rangle_{\rm c}, \tag{6.52}$$

whence

$$\begin{aligned} \mathrm{d} \ln Z^{\rm c} &= \left.\frac{\partial \ln Z^{\rm c}}{\partial \alpha_k}\right)_{\boldsymbol{\lambda}^{\rm c}} \mathrm{d}\alpha_k + \left.\frac{\partial \ln Z^{\rm c}}{\partial \boldsymbol{\lambda}^{\rm c}}\right)_{\boldsymbol{\alpha}} \cdot \mathbf{d}\boldsymbol{\lambda}^{\rm c} && (6.53) \\ &\quad \text{(summation on } k) \\ &= -\boldsymbol{\lambda}^{\rm c} \cdot \langle \mathbf{d}\boldsymbol{f} \rangle_{\rm c} - \boldsymbol{F} \cdot \mathbf{d}\boldsymbol{\lambda}^{\rm c}. && (6.54) \end{aligned}$$

From (6.51),

$$\begin{aligned} \mathrm{d}S^{\rm c} &= \mathrm{d} \ln Z^{\rm c} + \mathbf{d}\boldsymbol{\lambda}^{\rm c} \cdot \langle \boldsymbol{f} \rangle_{\rm c} + \boldsymbol{\lambda}^{\rm c} \cdot \mathbf{d}\langle \boldsymbol{f} \rangle_{\rm c} && (6.55) \\ &= \boldsymbol{\lambda}^{\rm c} \cdot (\mathbf{d}\langle \boldsymbol{f} \rangle_{\rm c} - \langle \mathbf{d}\boldsymbol{f} \rangle_{\rm c}). && (6.56) \end{aligned}$$

Define the inexact differential

$$\begin{aligned} \text{đ}\boldsymbol{Q} &\equiv \mathbf{d}\langle \boldsymbol{f} \rangle_{\rm c} - \langle \mathbf{d}\boldsymbol{f} \rangle_{\rm c} && (6.57) \\ &= \sum_i \boldsymbol{f}(i, \boldsymbol{\alpha})\, \mathrm{d}p^{\rm c}(i, \boldsymbol{\alpha}), && (6.58) \end{aligned}$$

so that

$$\mathrm{d}S^{\rm c} = \boldsymbol{\lambda}^{\rm c} \cdot \text{đ}\boldsymbol{Q}. \tag{6.59}$$

The Lagrange multipliers act as integrating factors. If $\text{đ}Q_{j=j'}$ is zero, so that $\mathrm{d}\langle f_{j'} \rangle_{\rm c} = \langle \mathrm{d} f_{j'} \rangle_{\rm c}$, the constraint variable $F_{j'} \equiv \langle f_{j'} \rangle_{\rm c}$ *follows* the constraint function $f_{j'}$ through the change ('adiabatically').

Next,

$$\begin{aligned} \mathrm{d}p^{\rm c}(i, \boldsymbol{\alpha}) &= \mathrm{d}\big(Z^{\rm c\,-1} \exp(-\boldsymbol{\lambda}^{\rm c} \cdot \boldsymbol{f}) \big) && (6.60) \\ &= -Z^{\rm c\,-1} \exp(-\boldsymbol{\lambda}^{\rm c} \cdot \boldsymbol{f})\, \mathrm{d}(\ln Z^{\rm c} + \boldsymbol{\lambda}^{\rm c} \cdot \boldsymbol{f}) && (6.61) \end{aligned}$$

so that

$$\begin{aligned} \langle \mathrm{d}p^{\rm c}(i, \boldsymbol{\alpha}) \rangle_{\rm c} &= -\mathrm{d} \ln Z^{\rm c} - \langle \mathrm{d}(\boldsymbol{\lambda}^{\rm c} \cdot \boldsymbol{f}) \rangle_{\rm c} && (6.62) \\ &= 0 && (6.63) \end{aligned}$$

in view of (6.54).

The variation in the maximized entropy (6.56) can be rewritten as

$$\mathrm{d}S^{\rm c}[\boldsymbol{F}, \boldsymbol{\alpha}] = \boldsymbol{\lambda}^{\rm c} \cdot \mathbf{d}\boldsymbol{F} - \boldsymbol{\lambda}^{\rm c} \cdot \left\langle \frac{\partial \boldsymbol{f}}{\partial \alpha_k} \right\rangle_{\rm c} \mathrm{d}\alpha_k, \tag{6.64}$$

whence

$$\left.\frac{\partial S^{\mathrm{c}}}{\partial \alpha_k}\right)_{\boldsymbol{F}} = -\boldsymbol{\lambda}^{\mathrm{c}} \cdot \left\langle \frac{\partial \boldsymbol{f}}{\partial \alpha_k} \right\rangle_{\mathrm{c}} \tag{6.65}$$

$$= \left.\frac{\partial \ln Z^{\mathrm{c}}}{\partial \alpha_k}\right)_{\boldsymbol{\lambda}^{\mathrm{c}}}, \tag{6.66}$$

a noteworthy relation. Now compute, for any function $\phi(i, \boldsymbol{\alpha})$, the quantity

$$\mathrm{d}\langle\phi\rangle_{\mathrm{c}} - \langle \mathrm{d}\phi\rangle_{\mathrm{c}} = \sum_i \phi(i, \boldsymbol{\alpha})\, \mathrm{d}p^{\mathrm{c}}(i, \boldsymbol{\alpha}). \tag{6.67}$$

Substitution of (6.61) into this, followed by (6.54), gives the result

$$\mathrm{d}\langle\phi\rangle_{\mathrm{c}} - \langle \mathrm{d}\phi\rangle_{\mathrm{c}} = -\boldsymbol{\lambda}^{\mathrm{c}} \cdot \big(\langle \phi\, \mathbf{d}\boldsymbol{f}\rangle_{\mathrm{c}} - \langle\phi\rangle_{\mathrm{c}}\langle \mathbf{d}\boldsymbol{f}\rangle_{\mathrm{c}}\big) - \mathbf{d}\boldsymbol{\lambda}^{\mathrm{c}} \cdot \big(\langle \phi\, \boldsymbol{f}\rangle_{\mathrm{c}} - \langle\phi\rangle_{\mathrm{c}}\langle \boldsymbol{f}\rangle_{\mathrm{c}}\big). \tag{6.68}$$

This is the basis of many useful relations. For example one may take $\phi = \boldsymbol{f}$ or $\partial\boldsymbol{f}/\partial\boldsymbol{\alpha}$, and divide by $\mathbf{d}\boldsymbol{\alpha}$ while holding $\boldsymbol{\lambda}^{\mathrm{c}}$ constant (pure mathematicians will forgive the phrasing), to give a covariance relation among the α-derivatives of the constraint functions. Further covariance relations are available by differentiating $\ln Z^{\mathrm{c}}$.

This entire formalism is an exercise in probability assignment. Those who have seen it in a different context are asked not to think ahead.

6.4 Predictive statistical mechanics

The programme for statistical mechanics is set as an exercise in inference. Consider the example of finding the pressure within a gas. Pressure at a restraining wall is to do with the rate at which the gas transfers momentum to unit area of the wall. At the atomic level of description this varies at all times; but fine details of the microstate should not affect the pressure, for the notion of pressure predated the notion of microstate. An **averaging** must be involved in the definition of pressure (and other macrovariables). In any case we are not certain as to the initial microstate: we assign a probability distribution over microstates according to our prior *macro*knowledge; and pressure is appropriately defined as the expectation, over this distribution, of momentum flux. Certain other macrovariables are likewise expectations of microfunctions defined on Γ, so that probability assignment and macroscopic prediction proceeds according to the canonical formalism of Section 6.3. The remaining macrovariables are the α_k upon which depend the microfunctions: quantities such as volume or electric field. These variables may be extensive or intensive.

If the measured pressure differs from the predicted value, then the conditioning information has proved insufficient for macroscopic prediction.

In systems of $\sim 10^{23}$ particles this invariably means that a macrovariable relevant for macroscopic inference has been ignored; for example, the bulk dipole moment of a polar gas in a changing electric field. There is everything to gain by making macroscopic prediction and then comparing with observation: agreement means that a closed set of macrovariables for the situation has been found, while disagreement means that a macrovariable of relevance is missing, and can subsequently be sought.

Discrepancy might in principle be due instead to our having alighted on a rogue microstate consistent with the macroinformation but in which, say, all the particles move to one end of the container. Intuitively, such rogues should be encountered ever more rarely as the number of particles increases, and in fact there are enough particles in typical macroscopic systems to make it worthwhile taking macroscopic reproducibility as an *axiom*. Our formalism cannot then be amended to treat probabilities of macrovariables, since the 'probability of a probability' is without meaning. Should probabilities of macrovariables be desired, dynamical definitions must be used; the individual microstates consistent with the experimental data must be tracked; and irrelevancies marginalized out *a posteriori*. While this may be possible in computer simulations of a few tens or even hundreds of particles, it is inconceivable for 10^{23} of them.

Next, we examine what it is meant by the number of microstates consistent with a macrostate. The probability distribution on Γ, which is assigned according to the macrostate, is identified as the Liouville function ρ, and the information entropy, which is to be maximized, as the Gibbs entropy (6.6). (The measure on phase space is uniform, by symmetry between coordinates and conjugate momenta.) The abundance of consistent microstates is now characterized by the magnitude W of the phase volume Σ enclosed by a surface of constant probability, within which the probability of finding the microstate is close to unity:

$$\begin{aligned} \int_\Sigma \mathrm{d}\Gamma\, \rho &= 1-\epsilon \qquad \text{with } \rho = \text{constant on } \partial\Sigma \\ \text{and} \qquad W(\epsilon) &= \int_\Sigma \mathrm{d}\Gamma. \end{aligned} \tag{6.69}$$

But how close? What value of ϵ is appropriate? Remarkably this does not matter to macroscopic prediction in the limit of very many degrees ($6N$) of freedom, or many particles (N). This is the **asymptotic equipartition theorem** of Shannon (1948): provided that the maximum entropy distribution predicts relative energy fluctuations vanishing as $N^{-1/2}$ in the limit, and that the physical meanings of the constraints remain constant in this limit, then S^{c}/N, the maximized information entropy per particle, and $(\ln W(\epsilon))/N$ approach the same definite limit irrespective of the value of ϵ as long as $\epsilon \neq 0$ or 1. This is made clearer by understanding that W

is implicitly a function of the dimension of the phase space, $6N$: changing ϵ does of course change $W(N, \epsilon)$ dramatically, but the induced change in $\ln W(N, \epsilon)$ grows less rapidly than N and in the limit is swamped. This theorem is the explanation of Boltzmann's epitaph (6.5), which is extremely accurate for $N \approx 10^{23}$; and it implies that the canonical distribution, which maximizes S_G, defines the largest phase volume of reasonable probability over all distributions consistent with the constraints. As is usual with probability assignments, the idea is to stay as non-committal as possible.

The more constraints one has, the smaller is the maximized entropy and the 'likely phase volume'. Tiny changes in macrostate change this volume by factors of order $\exp(10^{23})$, and the enormity of this number underpins the great reliability of prediction within closed sets of macrovariables.

The intuitive meaning of the second law of thermodynamics was given in the Introduction: it is a consequence of the requirement that experiments on macroscopic variables be reproducible although different microstates are selected in each run. The Gibbs entropy S_G remains constant throughout any run. Its maximized value S^c, which is a function of the constraint variables—the macrostate—is the thermodynamic entropy. This latter increases; for example, let work be done on a system so as to raise its temperature from T_i at time $t_i = -\tau$ to T_f at time $t_f = 0$. The suffixes stand for *initial* and *final* and are not dummies. (Either state may be intermediate in a longer look at the system.) We have learned that T_f is reproducibly predicted given only three quantities: T_i, the energy supplied, and the heat capacity. Everything else—all 6×10^{23} bits of microdata—are irrelevant to the purpose. Denote by H the Hamiltonian of the system before heating, and by H' the auxiliary heating term; although this may not in practice be known, it *exists*. We assign an initial distribution according to the constraint that the expectation of H is the macroscopic energy U:

$$\langle H \rangle \equiv \int d\Gamma \, \rho H = U. \tag{6.70}$$

The canonical distribution (*not* ensemble) is

$$\rho_i^c = Z_i^{c-1} e^{-\beta_i^c H} \tag{6.71}$$

where

$$Z_i^c \equiv Z(\beta_i^c) = \int d\Gamma \, e^{-\beta_i^c H} \tag{6.72}$$

and

$$-\frac{\partial}{\partial \beta_i^c} \ln Z(\beta_i^c) = U_i, \tag{6.73}$$

with maximized entropy

$$S_i^c = \ln Z_i^c + \beta_i^c U_i \tag{6.74}$$

$$= -\beta_i^{c2} \frac{\partial}{\partial \beta_i^c} \left(\beta_i^{c-1} \ln Z(\beta_i^c) \right). \tag{6.75}$$

Historically, (6.71)–(6.75) was the first maximum entropy assignment, due to Boltzmann (1868, 1871*a*,*b*,*c*), who justified it combinatorially. The Lagrange multiplier β^{c} conjugate to thermodynamic energy is identified, by comparison of (6.71)–(6.74) with the formulae of thermodynamics, as the reciprocal of temperature:

$$\beta^{\mathrm{c}} = \frac{1}{T}. \tag{6.76}$$

In practice, it is temperature rather than energy which is specified. Since the two have a one-to-one relationship, specifying either leads to the same distribution. The inexact differential $đ\boldsymbol{Q}$ represents heat supplied irreversibly to the system. The third law of thermodynamics, concerning the limiting behaviour of S^{c} at low temperatures, can be studied from (6.75).

One might naïvely expect (6.71) to evolve during the heating according to Liouville's equation,

$$\frac{\partial \rho}{\partial t} + \left[\rho, H + H'\right]_{q,p} = 0 \tag{6.77}$$

to a distribution of canonical form but at the final inverse temperature β_{f}:

$$\rho_{\mathrm{f}}^{\mathrm{c}} \equiv Z_{\mathrm{f}}^{\mathrm{c}\,-1} \mathrm{e}^{-\beta_{\mathrm{f}} H}. \tag{6.78}$$

But this is trivial to disprove: Liouville's equation preserves the Gibbs entropy S_{G}, while $\rho_{\mathrm{i}}^{\mathrm{c}}$ and $\rho_{\mathrm{f}}^{\mathrm{c}}$ have differing Gibbs entropies. Denote the Liouville-evolute of $\rho_{\mathrm{i}}^{\mathrm{c}}$ by $\rho_{\mathrm{i}}(0)$: then $S_{\mathrm{G}}[\rho_{\mathrm{i}}(0)] = S_{\mathrm{G}}[\rho_{\mathrm{i}}^{\mathrm{c}}]$. But we require also that $\rho_{\mathrm{i}}(0)$ and $\rho_{\mathrm{f}}^{\mathrm{c}}$ correspond to the same macrostate, and since $\rho_{\mathrm{f}}^{\mathrm{c}}$ is chosen out of all such distributions to have maximum Gibbs entropy, it follows that $S_{\mathrm{G}}[\rho_{\mathrm{f}}^{\mathrm{c}}] \geq S_{\mathrm{G}}[\rho_{\mathrm{i}}(0)]$. Putting these relations together gives $S_{\mathrm{G}}[\rho_{\mathrm{f}}^{\mathrm{c}}] \geq S_{\mathrm{G}}[\rho_{\mathrm{i}}^{\mathrm{c}}]$, or $S_{\mathrm{f}}^{\mathrm{c}} \geq S_{\mathrm{i}}^{\mathrm{c}}$, a condition relating macrovariables. (Inequality is strict unless $T_{\mathrm{f}} = T_{\mathrm{i}}$.) This exemplifies the law of entropy increase. The failure of Liouville's equation to evolve $\rho_{\mathrm{i}}^{\mathrm{c}}$ into $\rho_{\mathrm{f}}^{\mathrm{c}}$ is not shocking once it is understood that ρ is not a physical property of the system, but a probability distribution and a predictive tool: a viewpoint accessible only to Bayesians. $H'(\Gamma)$ and $\rho_{\mathrm{i}}(0)$ contain an enormous amount of information irrelevant to macroscopic prediction, already pruned from $\rho_{\mathrm{f}}^{\mathrm{c}}$.

The maximum entropy formalism is obviously applicable to 'equilibrium' situations, in which the macrostate does not change with time; and where the α-variables (such as volume) change and the macrostate 'follows' through a sequence of equilibrium states in a thermodynamically reversible process. But the underlying notion of macroscopic reproducibility is as applicable to non-equilibrium as it is to equilibrium situations, and we can study the generalization to time-dependent constraints, furnishing a general formalism for non-equilibrium statistical mechanics. The value of a constraint variable at every instant throughout an interval $[-\tau, 0]$ constitutes an independent constraint:

$$\left\langle f(\Gamma; \boldsymbol{\alpha}(t)) \right\rangle = F(t) \quad \text{for} \quad -\tau \le t \le 0. \tag{6.79}$$

We use the functional calculus (see (6.47)), and maximize the Gibbs entropy (6.6) subject to normalization and (6.79), to give

$$\rho^{\mathrm{c}}(0) = \frac{1}{Z[\lambda^{\mathrm{c}}(t)]} \exp\left(- \int_{-\tau}^{0} \mathrm{d}t' \lambda^{\mathrm{c}}(t') f(\Gamma; \boldsymbol{\alpha}(t')) \right) \tag{6.80}$$

where

$$Z[\lambda^{\mathrm{c}}(t)] = \int \mathrm{d}\Gamma \exp\left(- \int_{-\tau}^{0} \mathrm{d}t' \lambda^{\mathrm{c}}(t') f(\Gamma; \boldsymbol{\alpha}(t')) \right) \tag{6.81}$$

and

$$-\frac{\delta}{\delta \lambda^{\mathrm{c}}(t)} \ln Z[\lambda^{\mathrm{c}}(\bullet)] = F(t) \qquad \text{in } -\tau \le t \le 0 \tag{6.82}$$

is a non-linear integral equation for $\lambda^{\mathrm{c}}(t)$. Prediction proceeds as usual for

$$\Phi(t) \equiv \int \mathrm{d}\Gamma\, \rho(t) \phi(\Gamma; \boldsymbol{\alpha}(t)) \tag{6.83}$$

both within $[-\tau, 0]$ and outside it. Propagation of $\rho^{\mathrm{c}}(0)$ to $\rho(t)$ is by means of an expansion in any forces tending to disturb canonical equilibrium; the results generalize today's linear response theory. This basis for non-equilibrium statistical mechanics, which transcends irreversibility, was first proposed by Jaynes (1963) and has been developed by Grandy (1988) for the quantum case, in which ρ is the density matrix; the formalism is closely related to the path integral version of quantum field theory. Spatial dependence is treated by further functional generalization: for t in (6.79)–(6.83) read $(t, \boldsymbol{r})$.

The maximized entropy $S^{\mathrm{c}}[F(t \in [-\tau, 0])]$ is related to the entropy obtained by continually maximizing S_{G} at every instant according to the instantaneous macrostate; the more nearly equal these are, the less relevant is the past, or future, to the present ('short memory'). $S^{\mathrm{c}}[F(t \in [-\tau, 0])]$ warrants a new name, and Jaynes has called it the **calibre**, because it measures the cross-section of a tube enclosing the world-lines in 'phase space-time' of the microstates which evolve consistently with the macro-information.

This formalism reproduces the fluctuation–dissipation theorem, and many other standard results of Kubo, Onsager, Einstein, Fokker and

Planck; and resolves disparities between them (Jaynes 1980; Grandy 1987, 1988). In particular, the equations of hydrodynamics, and expressions for transport coefficients in terms of interparticle potentials, have been derived (Grandy 1988). A brief critique of conventional kinetic theory from the present viewpoint is given in Appendix 6.C.

6.5 Ergodic theory is irrelevant

There is a notion that, despite the crucial role of probabilistic inference, statistical mechanics retains its basis essentially in Hamiltonian dynamics. The averaging implied in the definition of macroscopic variables must then be temporal and spatial ('coarse graining'). Temporal averaging involves the **ergodic hypothesis**, that the probability expectation value of a function defined on the phase space, its 'ensemble average', equals its dynamical time average. But ergodic theory, though a valuable tool in studying orbit stability in planetary systems and particle accelerators, is not the basis for statistical mechanics. For the ergodic hypothesis is not always true: there are non-trivial counter-examples. (And incidentally when it is true, it is very difficult to prove for even the simplest cases: global studies of Hamiltonian systems are notoriously demanding (Arnol'd and Avez 1968).) Where ergodicity is proved, the timescale involved may run from thousands to billions of years, which is far greater than the timescale of thermodynamic experiments; this amounts to an experimental disproof of the relevance of ergodicity. Moreover, the ergodic timescale *increases* with the complexity of the system, against expectations that prediction should improve with the number of particles.

Ergodic theory is no help to non-equilibrium statistical mechanics, which by definition involves quantities changing in time. It is the basis neither for equilibrium nor non-equilibrium theory. The correct foundation is the principle of maximum entropy.

6.6 Conclusions

A precise definition of entropy has been given arising out of the business of macroscopic prediction. Macroscopic variables such as pressure are regarded as expectation values of microscopic functions defined on the full phase space of the system, with given macroscopic variables treated as constraints on the probability distribution of microstates. This distribution is assigned using the consistency criterion of maximum (Gibbs) entropy. Once assigned, the distribution can be used for prediction of other macrovariables. Thermodynamic entropy is identified with the maximized Gibbs entropy and is a function of the macrostate, not the microstate. The canonical formalism and its direct offspring, equilibrium thermodynamics, have been recovered. Provided that all relevant macroscopic variables are identified, prediction is exceedingly reliable in consequence of the enormous number of particles in macroscopic systems.

The basis of thermodynamic entropy increase lies in recognition that microdescription and macrodescription are fundamentally distinct. As a video-camera zooms in or out there is no continuous transition from one to the other. Experimental results on macrovariables are required to be reproducible even though the microstate differs from run to run. This condition defines a set of microstates upon which the probability distribution of microstates is defined; thermodynamic entropy increase corresponds to increase of the size of this set whenever information on the macrohistory of the system is lost.

Unless it is appreciated that the probability distribution is not a physical property of the system but a tool of predictive inference, this programme cannot be comprehended. Consequently a detailed description of Bayesian probability theory, and its foundation in consistency principles, have been given. Since this and the meaning of entropy are both controversial, some differing viewpoints have been outlined. It has been shown how the present view resolves the problems within these.

It is emphasized that the notion of macroscopic reproducibility and the predictive framework of maximum entropy are equally applicable out of equilibrium. These concepts are therefore also the basis for non-equilibrium statistical mechanics, whose formalism has been outlined. Many known results have been rederived effortlessly from it, and the reader is now in a position to witness its further development (Grandy 1987, 1988).

6.A Appendix: Other views of probability

There are three principal schools of probability: that which holds that probability is an entirely personal construct, and that anyone can assign to anything any probability; that whose assignments are constrained by consistency (the present view); and that which holds that probability is a physical quantity—of which more shortly. Adherents to each school use 'subjective' and 'objective' differently in describing each other, and since three into two won't go, these terms are best avoided.

The first of these schools calls itself **personalistic** or **subjective Bayesian**, and it is inconsistent in the senses described in Section 6.2. The most influential view is the subset of the third school known as **frequentist**. This school asserts that probability is only meaningful as a relative frequency: the probability of heads in coin tossing is *defined* as the ratio of the number of heads to the number of tosses as the latter increases without bound. Further, only conditioning information in the form of frequencies is acceptable. There are problems with this: trials must be 'random', an unspecified concept; also one in practice never has an infinity of trials, so that in a big enough space the finite number recorded still comprise a single event. There is no problem for Bayesians in speaking of the probability of a single event—such as tomorrow's weather—and no problem either in deductive logic, which we have seen is a special case; but those frequentists who even admit the propriety of the idea are forced to invent a fictitious randomizing agent.

Neglect of prior information in other forms than frequencies represents a serious loss: one would hardly trust a doctor who stated that 70% of patients with a given symptom needed an operation, but who refused to examine medical histories of individuals. Non-frequency information can always be incorporated by embedding it in an imaginary 'ensemble' of events which might have happened (but didn't), but one needs to know the multiplicities of the differing events in the ensemble and these are calculable only from the prior information, again. If this is done correctly, the result coincides with the Bayesian; but the interpretation obscures the issue. Also, frequentists interpret marginalization as summing over a *real* ensemble even though there is only one system in one—unknown—state.

Frequentists assert that incorporation of non-frequency information places all Bayesians in the personalistic camp, and that there is a sharp distinction between 'statistical inference', which uses non-frequency information, and 'statistics', which supposedly does not. The frequency interpretation derives, historically, from gambling and other combinatorial problems in which each of a set of events is implicitly assigned equal prior probability (by symmetry), and today is defended on grounds of 'objectivity'; but this notion is already (and is only) captured by consistency conditions. A frequency is a physical observation used as conditioning information, and is not a probability. The frequency view is harmful in its attribution of physical meaning to fictional randomizing ('stochastic') processes; Jaynes (1989) has called this the **mind projection fallacy**. It is widespread in statistical mechanics, where clear distinction between inference and dynamics is essential.

There is a variety of *ad hoc* methods for solving statistical problems in which the data consist of repeated samples of a variable. That there are several ways of answering the same question is a weakness, not a strength, for only one way can be consistent and that one is based on the laws of probability and nothing else. Any other method can be made to look as silly as we like by suitable choice of data set. Most 'sampling-theoretical' methods were developed during the Bayesian decline of the late 19th and early 20th century for use in problems with data sets of particular sorts, often Gaussian, and the resulting answers were good—meaning, with hindsight, close to the Bayesian. This bears tribute to the original workers. Trouble arose when the next generation of texts set out these methods as completely general ones. Jaynes (1976) compares several of them to the

Bayesian, and the most striking comparison now follows.

A company manufactures electric bar fire elements. The process is such that elements in continuous use fail with constant probability per unit time, so that the expected number still in operation decays exponentially. The probability of failure of an element in a short interval at time t decays similarly: it is less likely to fail at large times because it already probably has failed. The time constant, determined from a large enough sample of elements, is such that $100(1 - e^{-1}) \approx 63.2\%$ have failed within one year. (The exponential form may be confirmed by continuing inspection of this sample.)

To extend their working life, the company coats each new element with an identical protective layer, which sublimates away at a constant but unknown rate. The elements suffer no degradation while protected, so that once the protector is gone they begin to fail as previously. Three customers state that their protected elements failed after $t_1 = 12$, $t_2 = 14$ and $t_3 = 16$ years of continuous use. What is the best estimate of the time θ taken for the protector to wear off?

We know with certainty that the answer is not more than 12 years, for failure occurs only after wear-away, and one element failed precisely 12 years after installation. The Bayesian calculation reproduces this feature (more accurately, the brain recognizes this feature of the Bayesian calculation): the sampling density $P(t|\theta)$, shown in Fig. 6.3, is the displaced exponential $H(t-\theta)\,e^{\theta-t}$ ($H(x)$ is zero for $x < 0$ and unity otherwise), and the likelihood is the product $P(t_1|\theta)\,P(t_2|\theta)\,P(t_3|\theta)$: this is how Bayesians handle repeated sampling. This product simplifies to $H(12-\theta)\,e^{3(\theta-14)}$, because the other step functions are redundant. The posterior density $P(\theta|\{t_i\})$ for θ, encapsulating the required result, is proportional to the likelihood and incorporates the cut-off; it also depends on the prior probability density for θ, based on whatever is known about the coating and wear-away processes.

A popular sampling-theoretical method for this problem is called **unbiased estimators**. An estimator is a symmetrical function of the data $\{t_i\}$ whose probability distribution is used in estimating the distribution of the desired quantity. One common estimator is linear in the data and is scaled to a single sample, and an **unbiased** estimator of θ is offset such that its expectation value equals θ. Here, the linear unbiased estimator is $\theta^* = \frac{1}{3}(t_1 + t_2 + t_3 - 3)$. A contour integration leads to the distribution

$$P(\theta^*|\theta)\mathrm{d}\theta^* = \tfrac{27}{2}H(1+\theta^*-\theta)(1+\theta^*-\theta)^2 e^{-3(1+\theta^*-\theta)}\mathrm{d}\theta^*. \qquad (6.84)$$

This is now supposed to be the distribution for θ: $P(\theta|\theta^*)$, in contravention of the product rule, and the impropriety of having any answer without specifying prior knowledge of θ (such as $\theta > 0$). It is cut off not at the smallest sample, 12, but at the sample mean, 14. It peaks at $\theta = 13\frac{1}{3}$;

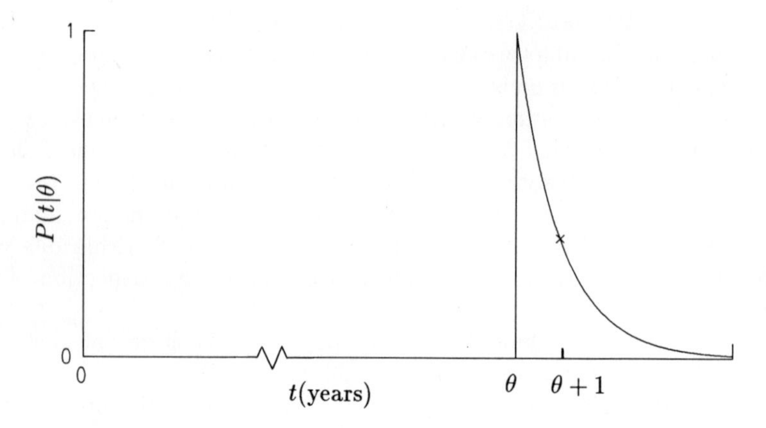

Fig. 6.3. The sampling density $P(t|\theta)$.

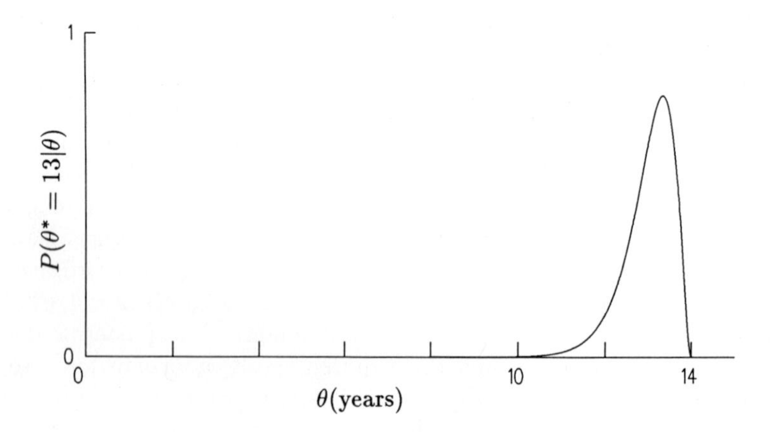

Fig. 6.4. The unbiased estimator density for θ.

and 93.8% of it lies, impossibly, above $\theta = 12$, with only a small tail in the allowed region (Fig. 6.4). This startling result strikingly exemplifies the inconsistency (and complexity) of using anything other than the sum and product rules. Such inconsistent methods of data reduction remain in routine use in fields such as particle physics, economics, industry and medicine.

We rebut some anticipated objections. The estimator θ^* is not a **sufficient statistic**—a single function of the data encapsulating everything relevant for estimating θ. But, no systematic way of constructing a sufficient statistic is known (there are some so-called Bayesian techniques, but this is a misnomer); the true Bayesian calculation has no need of any; and sufficient statistics simply do not exist in some problems, for example in the estimation of the upper and lower cut-offs of a 'top hat' distribution, for which the smallest and largest samples are *both* needed.

A second example concerns independent binary ('Bernoulli') trials. If it is accepted that the probability of heads in the next throw of a (possibly weighted) coin, given six heads in ten throws, is $\frac{6}{10}$, then it follows from the product rule that the probability of heads in the next two throws—the 11th and 12th—is $(6/10) \times (7/11)$. Yet many texts, which quote the product rule approvingly, give the answer as $(6/10) \times (6/10)$.

Only the Bayesian view is consistent.

6.B Appendix: Bayes' theorem; continuous spaces and measure; improper priors

The updating rule (6.11) is usefully rewritten in the form

$$\begin{aligned} p(X|Y,I) &= Kp(X|I)p(Y|X,I), \\ \text{with} \quad K^{-1} &= p(X|I)p(Y|X,I) + p(\bar{X}|I)p(Y|\bar{X},I). \end{aligned} \tag{6.85}$$

Of course K^{-1} is just $p(Y|I)$, expanded by means of the marginalizing rule; but here it obviously acts as a normalizing factor, and in real updating problems can be calculated *a posteriori*. (6.85) is then known as Bayes' theorem, with $p(X|I)$ the prior probability, $p(X|Y,I)$ the posterior probability, and $p(Y|X,I)$ the likelihood. This relation between the posterior and the likelihood is called the principle of inverse reasoning: for example, for the tax office to work out whether someone has an undeclared extra source of revenue, it examines how people with extra revenue live, and compares that with the lifestyle of the suspect. Here, p(extra revenue|extravagant lifestyle) and p(extravagant lifestyle|extra revenue) are mutually related. In Bayes' theorem, the posterior increases as the prior increases, though not in proportion because of the normalizing factor K in (6.85) taken together with the relation $p(\bar{X}|I) = 1 - p(X|I)$.

Bayes' theorem is central to hypothesis testing. Suppose, for example, that we wish to test the hypothesis that a metal bar is 50 cm long (proposition X), given that it was measured at 50 cm (proposition Y) in an experiment taking Gaussian error distribution about the true length with standard deviation 0.5 cm. The prior information I is that the bar looks roughly 50 cm by eye; assignment of priors from such diffuse information is covered shortly. Testing this hypothesis is a routine application of Bayes' theorem, taking the error density for the observations as likelihood. Of course a hypothesis cannot be tested in isolation, but rather against other hypotheses; its negation $\bar{X}$ consequently demands specification. Here, the hypothesis space consists of all lengths from zero to infinity.

Should the prior be broad, corresponding to large prior uncertainty, the peak in the posterior is dominated by the likelihood: the data. If the prior is sharp, so that we begin confidently, the peak in the posterior is dominated by the peak in the prior and discrepancy with the data is ascribed to experimental error. All this accords with intuition.

Scientific methodology itself is an example of hypothesis (or theory) testing. Within it is a precise expression of Ockham's razor principle, that the simplest theory to fit the facts closely is preferred. There is a trade-off between goodness of fit and simplicity of the theory: obviously a theory with more floating parameters can fit noisy data better than one in which the parameters are asserted to have particular values; but it pays a penalty in placing some of the prior probability for the parameters where the data and the noise statistics subsequently indicate these are very unlikely to be.

A remarkable illustration of the inferential role of the sum and product rules lies in Bell's demonstration of nonlocality (Bell 1964). Posit that the result of a spin measurement of a spin-$\frac{1}{2}$ fermion is governed by a variable internal to the particle (that is, local and 'hidden'). Now examine pairs of particles having zero combined angular momentum (singlet state), implying that their internal variables are correlated: information about the internal variable of one particle tells us something about the internal variable of the other. By measuring the spin of one particle along a particular direction, one is able to learn something about its internal variable prior to measurement, through an inverse problem. Through the correlation, one now infers something about the internal variable of the second particle, which may be arbitrarily distant from the first; locality has it that the first measurement has no effect on the internal variable of the second particle. From what we have inferred about the internal variable of the second particle we now make (probabilistic) predictions about the result of a spin measurement upon it. Lack of knowledge of the details of the internal variable theory induces an *inequality*, which is experimentally violated, and consequently locality—the only physical assumption involved—fails. The (quantum) physics of the particles is not involved, since the analysis is a chain of inference. Each step in it has a precise counterpart in the Bayesian analysis of the situation (Garrett 1990*a*,*b*).

What probability theory does not give are **adductive** rules for seeking which theories to test, or for deciding which parameters to let float. We call this **widening the argument**: the possibility space is augmented by taking the direct product of the old space with the space of each newly floating parameter. (The list of floating parameters is usefully termed the concept space. Its number of elements is the dimensionality of the possibility space.) A prior probability distribution is put on the floating parameters, and after the data are incorporated these parameters can, if desired, be marginalized out to give a distribution for other quantities which recognizes their relevance. Whenever the rules for widening the argument are found, the software side of the artificial intelligence effort will have been reduced to a matter of programming.

If the possibility space is (semi)infinite, the concept of the **improper prior**, which is non-normalizable, is often useful. This most usually arises on continuous spaces and so we discuss first the continuum, as a limit of the discretum (Jaynes 1963). Suppose that the number of points x_i with $i = 1, 2, \ldots, n$, in a discrete space, become ever more numerous, such that

$$\lim_{n\to\infty} \frac{\text{no. points in } \alpha \le x \le \beta}{n} = \int_\alpha^\beta \mathrm{d}x\, m(x) \qquad \forall \alpha, \beta. \tag{6.86}$$

Here, $m(x)$ is a (Lebesgue) measure, and

$$\lim_{n\to\infty} [n(x_{i+1} - x_i)] = 1/m(x_i). \tag{6.87}$$

The discrete probability distribution $p(i)$ becomes a probability density $P(x)$ according to

$$p(i) = P(x_i)(x_{i+1} - x_i) \tag{6.88}$$

so that

$$p(i) \to P(x_i)/\big(nm(x_i)\big) \tag{6.89}$$

and

$$\sum_i p(i)\ldots \to \int \mathrm{d}x\, P(x)\ldots. \tag{6.90}$$

Therefore the discrete information entropy goes over to

$$S \equiv -\sum_i p(i)\ln p(i) \to -\int \mathrm{d}x\, P(x) \ln\left(\frac{P(x)}{m(x)}\right). \tag{6.91}$$

There is also an additive term $\ln n$, which diverges in the limit. Since only entropy *differences* are important this term has been ignored, at a cost of destroying positive-definiteness. The resulting expression (6.91) has the essential property of invariance under changes of label: if we work not with x but with $y(x)$, then the probability density and the measure, both labelled by a subscript, transform according to

$$\begin{aligned} P_y(y)\mathrm{d}y &= P_x(x)\mathrm{d}x, \\ m_y(y)\mathrm{d}y &= m_x(x)\mathrm{d}x, \end{aligned} \tag{6.92}$$

so that

$$S_x = -\int \mathrm{d}x\, P_x(x)\ln\left(\frac{P_x(x)}{m_x(x)}\right) = -\int \mathrm{d}y\, P_y(y)\ln\left(\frac{P_y(y)}{m_y(y)}\right) = S_y. \tag{6.93}$$

In all of this we are assuming uniform weighting ('degeneracy') on the points which go over to the continuum; generalization is routine.

The measure is determined by symmetry considerations. In the discrete case, entropy maximization with no constraints other than normalization leads to a probability distribution equal at each point. This is demanded also by the symmetry ('indifference') of the problem: if we have no information favouring one point over another, we must assign equal probability to all. In the continuum case, unconstrained entropy maximization leads to the density $P(x) \propto m(x)$. Group-theoretical arguments, the formal expression of symmetry, can be used to assign the unconstrained $P(x)$, and hence $m(x)$. An example now illustrates the idea.

Let us assign the unconstrained probability density, so as to determine the measure for the location of a bead on a long wire. What is meant by saying we are *completely ignorant* of the location? The answer is that our state of knowledge, summarized in the distribution, is unchanged if we shift by some distance. No signposts are visible. Therefore, if x and x' represent two location variables, related by the shift $x' = x + \gamma$, we require that

$$P(x)\mathrm{d}x = P(x')\mathrm{d}x'. \tag{6.94}$$

The transformation from x to x' has Jacobian one, so that (6.94) reduces to the functional equation

$$P(x) = P(x + \gamma), \tag{6.95}$$

with unique solution $P(x) =$ constant, to be determined by normalization. Hence the measure $m(x)$ for this problem is a constant. Intuition gives the same result in this simple situation.

This 'transformation group' method is due to Jaynes (1968). While it may not be the last word in measure determination, it refutes the notion that continuum problems are ill-defined. 'Uninformative prior' does not mean 'no prior'.

If the wire is infinite, then although the measure is well-behaved, the unconstrained probability—which is a prior, if data are due—is non-normalizable, or **improper**. This does not matter in Bayes' theorem, (6.85), since the prior appears both on numerator and denominator so that its normalization is irrelevant.

Nevertheless a problem remains: a non-normalizable function cannot represent a probability or a state of knowledge, prior or otherwise. This is resolved on comprehending that there is additional testable information available from other sources before data samples come in, such as information concerning the mean or the standard deviation. When incorporated via maximum entropy, this will render the pre-data distribution normalizable. That such information is often imprecise is beside the point. For example, the luminosity of a star cannot be less than a certain value or we would not even know of the star's existence, and it cannot exceed a threshold value or sunglasses would be needed. Strictly speaking, this information should always be included; but if the product of the improper prior and the likelihood function for subsequent data is normalizable, so that the posterior is proper, the effect of ignoring the information is invariably negligible. And any experiment whose likelihood leads to an improper posterior would be far too uninformative ever to be proposed. The improper prior is, therefore, a useful labour-saving device.

6.C Appendix: Kinetic theory

Of the other schools of non-equilibrium theory we mention one, founded, ironically, by Boltzmann. The idea is that the Liouville function, which by indifference is symmetric in the canonical coordinates of all (identical) particles, is integrated over the coordinates of $N-1$ of them to give a function of a single pair (p_1, q_1), called the one-particle distribution function, f_1. This lies at the end of a chain in which $N-n$ particle coordinates are integrated to give the 'n-particle distribution function'. Partial integration of the Liouville equation generates recurrence relations for the evolution of the nth function in terms of the $(n+1)$th: the **BBGKY hierarchy** (Grandy 1987). A further, **closure** approximation is needed if the one-particle distribution function (or any other) is to determine its own future, and this programme and its consequences are called **kinetic theory**. The first attempt at closure, by Boltzmann, met with some pragmatic success, but its detailed justification has remained controversial.

A traditional indicator of progress has been monotonicity of the quantity

$$H \equiv \int \mathrm{d}\Gamma_1 \, f_1 \ln f_1, \tag{6.96}$$

where $\Gamma_1 = (p_1, q_1)$ is the one-particle phase space. The quantity $-H$ is known as the Boltzmann entropy (it should not be confused with $k \ln W$). But, while it increases monotonically for certain closure approximations, it is certainly not monotonic in reality: Jaynes (1971) demonstrates its decrease for the free expansion of O_2 gas at 160 K and 45 atmospheres. No physically realistic closure therefore corresponds to monotonicity; and on this and on grounds of arbitrariness we reject (6.96) as a worthwhile definition of entropy. Clearly the failure of monotonicity is related to interactions among the particles, since in their absence independent particles remain independent and the Boltzmann and Gibbs entropies can be proved equivalent. One system of N particles is not equivalent to N systems of one particle in the presence of interactions.

The conventional Chapman–Enskog derivation of transport coefficients and hydrodynamic equations is based on the assumptions that f_1 relaxes rapidly to a 'local thermal equilibrium', which decreases exponentially in particle energy in the co-moving frame with a spatially varying fall-off constant; and that this form then decays slowly, in the absence of external influences, to global equilibrium. These assumptions are arbitrary and unnecessary (Grandy 1988).

Despite these criticisms, the BBGKY hierarchy is sufficiently interesting that it should not be abandoned out of hand, and a complementary, Bayesian approach has recently been begun (Jaynes 1986).

References

Arnol'd, V. I. and Avez, A. (1968). *Ergodic problems of classical mechanics.* Benjamin, New York. (Reprinted (1989) by Addison-Wesley, Reading, Massachusetts.)

Bell, J. S. (1964). On the Einstein Podolsky Rosen paradox. *Physics*, **1**, 195–200.

Boltzmann, L. (1868). Studien über das Gleichgewicht der lebendige Kraft zwischen bewegten materiellen Punkten. *Sitzungsberichte der kaiserliche Akademie der Wissenschaften in Wien, Mathematisch-Naturwissenschaftliche Klasse IIa*, **58**, 517–60.

Boltzmann, L. (1871*a*). Über das Wärmegleichgewicht zwischen mehratomigen Gasmolekülen. *Sitzungsberichte der kaiserliche Akademie der Wissenschaften in Wien, Mathematisch-Naturwissenschaftliche Klasse IIa*, **63**, 397–418.

Boltzmann, L. (1871*b*). Einige allgemeine Sätze über Wärmegleichgewicht. *Sitzungsberichte der kaiserliche Akademie der Wissenschaften in Wien, Mathematisch-Naturwissenschaftliche Klasse IIa*, **63**, 679–711.

Boltzmann, L. (1871*c*). Analytischer Beweis des zweiten Hauptsatzes der mechanischen Wärmtheorie aus den Sätzen über das Gleichgewicht der lebendigen Kraft. *Sitzungsberichte der kaiserliche Akademie der Wissenschaften in Wien, Mathematisch-Naturwissenschaftliche Klasse IIa*, **63**, 712–32.

Boltzmann, L. (1909). *Wissenschaften Abhandlungen I* (ed. F. Hasenöhrl), pp. 49–96, 237–58, 259–87, 288–308. Leipzig. Reprints of (Boltzmann 1868, 1871*a*,*b*,*c*).

Brush, S. G. (1983). *Statistical physics and the atomic theory of matter, from Boyle and Newton to Landau and Onsager.* Princeton University Press.

Cox, R. T. (1946). Probability, frequency and reasonable expectation. *American Journal of Physics*, **14**, 1–13.

Garrett, A. J. M. (1990*a*). Bell's theorem, inference, and quantum transactions. *Foundations of Physics*, **20**, 381–402.

Garrett, A. J. M. (1990*b*). Bell's theorem and Bayes' theorem. *Foundations of Physics*, **20**, 1475–1512.

Gibbs, J. W. (1928). *The collected works of J. Willard Gibbs,* Vol. 1 and 2. (Including relevant papers published between 1873 and 1902). Longmans, Green and Co., New York. Reprinted Yale University Press (1936, 1948) and Dover, New York (1961).

Goldstein, H. (1980). *Classical mechanics* (2nd edn). Addison-Wesley, Reading, Massachusetts.

Grandy, W. T. Jr (1987). *Foundations of statistical mechanics, Volume 1: equilibrium theory.* Reidel, Dordrecht.

Grandy, W. T. Jr (1988). *Foundations of statistical mechanics, Volume 2: nonequilibrium phenomena.* Reidel, Dordrecht.

Jaynes, E. T. (1959). *Probability theory in science and engineering,* Colloquium lectures in pure and applied science, Vol. 4. Socony-Mobil Oil Company, Texas.

Jaynes, E. T. (1963). Information theory and statistical mechanics. In *Statistical physics,* 1962 Brandeis Lectures (ed. K. W. Ford), pp. 181–218. Benjamin, New York. Reprinted in Jaynes (1983).

Jaynes, E. T. (1968). Prior probabilities. *IEEE Transactions on Systems Science and Cybernetics*, **SSC-4** (Sept.), 227–41. Reprinted in Jaynes (1983).

Jaynes, E. T. (1971). Violation of Boltzmann's *H* theorem in real gases. *Physical Review A*, **4**, 747–50.

Jaynes, E. T. (1976). Confidence intervals vs Bayesian intervals. In *Foundations of probability theory, statistical inference and statistical theories of science* (ed. W. L. Harper and C. A. Hooker), pp. 175–257. Reidel, Dordrecht. Largely reprinted in Jaynes (1983).

Jaynes, E. T. (1980). The minimum entropy production principle. In *Annual Review of Physical Chemistry,* Vol. 31 (ed. S. Rabinovitch), pp. 579–601. Annual Reviews Inc., Palo Alto, California. Reprinted in Jaynes (1983).

Jaynes, E. T. (1983). *Papers on probability, statistics and statistical physics,* Synthese Library, Vol. 158 (ed. R. D. Rosenkrantz). Reidel, Dordrecht.

Jaynes, E. T. (1986). Some applications and extensions of the de Finetti representation theorem. In *Bayesian inference and decision techniques: essays in honour of Bruno de Finetti,* Studies in Bayesian econometrics, Vol. 6 (ed. P. Goel and A. Zellner), pp. 31–42. Elsevier, Amsterdam.

Jaynes, E. T. (1989). Clearing up mysteries—the original goal. In *Maximum entropy and Bayesian methods, Cambridge, England, 1988* (ed. J. Skilling), pp. 1–27. Kluwer, Dordrecht.

Laplace, Pierre-Simon, Marquis de (1820). *Théorie analytique des probabilités* (3rd edn). Courcier, Paris.

Shannon, C. E. (1948). A mathematical theory of communication. *Bell System Technical Journal*, **27**, 379–423 and 623–59. Reprinted in C. E.

Shannon and W. Weaver (1949). *The mathematical theory of communication.* University of Illinois Press, Urbana.

Tribus, M. (1969). *Rational descriptions, decisions and designs.* Pergamon, New York.

7

Some misconceptions about entropy

S. F. Gull

Abstract

We explore some misconceptions about statistical mechanics that are, unfortunately, still current in undergraduate physics teaching. The power of the Gibbs ensemble is emphasized and explained, and the second law of thermodynamics is proved, following Jaynes (1965). We then study the crucial role of information in irreversible processes and demonstrate by means of a concrete example how time-dependent data enable the *equilibrium* Gibbs ensemble to predict time-varying fluxes during the return to equilibrium.

7.1 Introduction

This contribution is the direct result of a discussion with second-year undergraduates at Cambridge that took place during a thermodynamics supervision. I asked them what they knew about entropy and about the statistical rationale for the second law. By way of answer they showed me their lecture notes, which reproduced *that awful H-theorem* given on page 39 of Waldram (1985). The main thrust of this chapter is to provide the antidote to that awful H-theorem and to draw attention to the beautiful proof of the second law given by Jaynes (1965). This short paper *'Gibbs vs Boltzmann entropies'* is a true masterpiece which states more clearly than anywhere else the key to the success of Gibbs' maximum entropy method.

> The Gibbs entropy of the canonical ensemble is *numerically* equal to the experimental entropy defined by Clausius.

If more physicists knew this simple, yet astonishing, fact the subject of thermodynamics (and plasma physics) would be far more advanced than it is today.

Jaynes's later work (1979, 1983) (see also Grandy (1987)) on non-equilibrium statistical mechanics is perhaps even more astounding, for it seems that Gibbs' maximum entropy algorithm is the complete answer for irreversible processes as well. Rather than discuss the general principles here, I illustrate this claim by an example: Brownian motion in one dimension. However, this little example is not nearly as trivial as it appears—there are some very serious lessons which can be learnt from it concerning non-equilibrium statistical mechanics.

Although it will not be mentioned again, it will become obvious that by championing the Gibbs/Jaynes/MaxEnt view of statistical mechanics, I am implicitly rejecting wholesale the approach of the 'Brussels' school (and indeed many other approaches). That implication is true: I do indeed feel that only the MaxEnt viewpoint distinguishes correctly the inferential content of statistical physics and separates it clearly from dynamical aspects. I believe that other approaches contain many misconceptions about the nature of inference in science. My title is, however, slightly ambiguous, and readers will have to judge for themselves on whose side the misconceptions lie.

7.2 Entropy in thermodynamics and statistical mechanics

The science of thermodynamics came of age when the concept of entropy was defined as a state variable for systems in thermal equilibrium. Although more modern and more erudite definitions exist, for our present needs we can restrict ourselves to the definition of the experimental entropy S_{E} in the form given in 1850 by Clausius (for a detailed account of the history of thermodynamics see Grandy (1987, Vol. 1, Appendix A))

$$\Delta S_{\mathrm{E}} = \int_{\substack{\text{reversible} \\ \text{path}}} \frac{\mathrm{d}Q}{T}, \tag{7.1}$$

where T is the absolute temperature and $\mathrm{d}Q$ is the amount of heat entering the system. In this way entropy is defined as a function of the macroscopic variables such as pressure and temperature, and its numerical value can be measured experimentally (up to a constant). This constant is provided for us by the third law, that S_{E} vanishes at the absolute zero of temperature.

Classical thermodynamics is the result of this macroscopic definition: it is conceptually clear and easy to apply in practice (I say this despite the effect it usually has on physics undergraduates!). Conceptual problems have arisen, though, when trying to give a microscopic, statistical interpretation. Statistical thermodynamics began in 1866 with Boltzmann's kinetic theory of gases. He considered a gas of N particles each in a 6-dimensional phase space of position and momentum, and studied how collisions led to an equilibrium distribution. He defined an H function, which we relate to a Boltzmann entropy $S_{\mathrm{B}} \equiv -k_{\mathrm{B}} H$,

$$H = \int \mathrm{d}^3x \, \mathrm{d}^3p \, \rho \log \rho, \tag{7.2}$$

where $\rho(x, p, t)$ is the distribution of particles.

A little later, the statistical mechanics of Gibbs was developed. Gibbs focused attention on the $6N$-dimensional joint phase space of the N particles, and to introduce statistical notions he employed the artifice of the *ensemble*, a large number of copies of the system. These (imagined) copies of the system provided insight into what the actual system might be doing.

Although many of the early results are due to Boltzmann, it was Gibbs who gave us the basic tool of statistical thermodynamics: the Gibbs algorithm. In order to set up the equilibrium ensemble, we maximize the Gibbs entropy

$$S_{\mathrm{G}} = -k_{\mathrm{B}} \int \mathrm{d}\tau \, p_N \log p_N \tag{7.3}$$

under the available constraints (e.g., the ensemble average energy $\langle E \rangle \equiv \int \mathrm{d}\tau \, E p_N$), where p_N is the probability density function (p.d.f.) for the N-particle system. This method is successful to this day. Even the transition to quantum mechanics passed without incident: the quantum mechanical definition involves the density matrix $S_{\mathrm{G}} = -k_{\mathrm{B}} \mathrm{Trace}(\rho \log \rho)$.

The situation in physics teaching today is still unsatisfactory, despite the everyday success of statistical mechanics with practical problems in physics and chemistry. We use the Gibbs algorithm in any detailed calculation, but teachers often try to justify this using the language of Boltzmann. As a result of this mixture of ideas, there are misconceptions about statistical mechanics. These misconceptions stem from a basic misunderstanding about the role of probability theory in physics, and it is there we must start.

7.3 Inference: the ground rules

I now give, without apology, a modern-day Bayesian viewpoint of the nature of inductive inference.

7.3.1 Bayes' theorem

In its simplest form this elementary theorem relates the probabilities of two events or hypotheses A and B. It states that the joint probability distribution function (p.d.f.) of A and B can be expressed in terms of the marginal and conditional distributions:

$$\mathrm{pr}(A, B) = \mathrm{pr}(A)\mathrm{pr}(B|A) = \mathrm{pr}(B)\mathrm{pr}(A|B). \tag{7.4}$$

Bayes' theorem is merely a rearrangement of this decomposition, which itself follows from the requirement of consistency in the manipulations of probability (Cox 1946). Although anyone can prove this theorem, those who believe it and use it are called Bayesians. Before using it, however, the joint p.d.f. has to be assigned. Because Bayes' theorem is simply a rule for manipulating probabilities, it cannot by itself help us to assign them in the first place, and for that we have to look elsewhere.

7.3.2 Maximum entropy

The maximum entropy principle (MaxEnt) is a variational principle for the assignment of probabilities under certain types of constraint called *testable information*. Such constraints refer to the probability distribution directly: e.g., for a discrete p.d.f. $\{p_i\}$, the ensemble average $\langle r\rangle \equiv \sum_i r_i p_i$ of a quantity r constitutes testable information. MaxEnt states that the probabilities should be assigned by maximizing the entropy

$$S = -\sum_i p_i \log \frac{p_i}{m_i}, \tag{7.5}$$

under the constraints $\sum_i p_i = 1$ and $\langle r\rangle = r_0$, where $\{m_i\}$ is a suitable measure over the space of possibilities (hypothesis space). The MaxEnt rule can be justified as the only consistent variational principle for the assignment of probability distributions (Shore and Johnson 1980; Gull and Skilling 1984). It can also be justified in numerous other ways (Jaynes 1982). In the simplest case there is no additional information other than normalization: MaxEnt then gives equal probabilities to all possible events, in accordance with Bernoulli's *principle of insufficient reason*. In fact, I believe MaxEnt to be the only logical method we have for the assignment of probabilities—it is so powerful that it is all we need. MaxEnt is, of course, a rule for assigning probabilities once the hypothesis space has been defined: to choose the hypothesis space we again have to look elsewhere.

7.3.3 Inference and statistical mechanics

The real art is to choose an appropriate space of possibilities, and this is our task as physicists. At this level we enumerate the possible states of the system and investigate its dynamics. Indeed, most physicists work entirely at this level, studying dynamics. One could even say that the process of building models for systems constitutes 'real physics', at a philosophical level that we call ontology:

$$\text{models for reality} \equiv \text{ontology}.$$

Statistical mechanics, on the other hand, works almost entirely at the level of inference, where we are concerned with what we know about the state of the system:

knowledge about reality ≡ epistemology.

Seen this way, we realize that the Gibbs ensemble represents the probability that our N-particle system is in a particular microstate. In Gibbs' statistical mechanics we are making inferences about the state of a system, given incomplete information. We know the values of the macroscopic variables, but there are many microstates compatible with this macrostate. We are not assuming that the system actually explores all the states accessible within the constraints, or indeed that it changes state at all. When this is realized, we see that ergodic assumptions are irrelevant to the rationale of statistical mechanics, even if such theorems could be proved. Rather, we set up a probability distribution (ensemble) using MaxEnt and whatever constraints are available, and see what predictions result. If this process leads to experimentally verified predictions, well and good; it then follows that our information that led to this ensemble was sufficiently accurate and detailed enough for our purposes. If our predictions are not verified, we conclude that there must be other, unknown influences which are relevant and which should be sought at the ontological level.

7.4 Gibbs versus Boltzmann entropies

For the present purposes a single specific example will suffice to illustrate the difference between Boltzmann's kinetic theory and Gibbs's statistical mechanics. Suppose we consider a system of N interacting particles in a box of volume V, with a purely classical Hamiltonian

$$H = \sum_{i=1}^{N} \frac{\boldsymbol{p}_i^2}{2m} + U(\boldsymbol{x}_1, \boldsymbol{x}_2, \ldots, \boldsymbol{x}_N). \tag{7.6}$$

The Gibbs entropy S_{G} is defined in terms of the joint probability distribution p_N of the N particles,

$$S_{\mathrm{G}} = -k_{\mathrm{B}} \int \mathrm{d}\tau_N \, p_N \log p_N. \tag{7.7}$$

The Boltzmann distribution function requires a little reinterpretation, but we can make sense of it in terms of the single-particle distribution, defined as a marginal distribution over the N particles,

$$p_1 \equiv \int \mathrm{d}\tau_{N-1} \, p_N, \tag{7.8}$$

where the integration is over all particle coordinates except the first. The Boltzmann entropy $S_{\rm B}$ then becomes

$$S_{\rm B} = -k_{\rm B}N \int {\rm d}\tau_1\, p_1 \log p_1. \tag{7.9}$$

Of these two expression for the entropy, it should be immediately apparent that only Gibbs's definition is meaningful. No matter how big your system becomes, you always have one system with N particles in it, and not N systems each with one particle! However, the real power of Gibbs's definition lies in the following theorem proved by Jaynes (1965), which deserves to be more widely known. If the initial probability distribution of the system is that of maximum entropy $S_{\rm G}$ (the canonical ensemble), and the state variables are then altered over a locus of equilibrium states (reversible paths), then

$$\Delta S_{\rm G} = \int \frac{{\rm d}Q}{T}, \tag{7.10}$$

whereas

$$\Delta S_{\rm B} = \int \frac{{\rm d}\langle K\rangle + p_0\,{\rm d}V}{T}, \tag{7.11}$$

where Q is the heat input, K the kinetic energy, T the temperature and $p_0 \equiv Nk_{\rm B}T/V$, the equivalent pressure of a perfect gas. Hence the Gibbs entropy when maximized (i.e., for the canonical ensemble) can be identified *numerically* with the thermodynamic entropy defined by Clausius. More generally, because $S_{\rm G}$ is defined for all probability distributions, not just the canonical ensemble, we have

$$S_{\rm G} \leq S_{\rm E}, \tag{7.12}$$

with equality if and only if the distribution p_N is canonical.

On the other hand, the expression for the change of the Boltzmann entropy shows that it ignores both the internal energy and the effect of the inter-particle forces on the pressure. Because it is defined in terms of the single-particle distribution, it is difficult to see how the situation could be otherwise. The Boltzmann entropy is the same as the Clausius experimental entropy only for the case of a perfect gas, when it is equal to the maximized Gibbs entropy as well.

Our moral is simple: the Gibbs entropy is the correct theoretical concept because, when maximized, it is numerically equal to the experimental entropy. The Boltzmann entropy has no theoretical justification and is not equal to the experimental entropy.

7.4.1 That awful H-theorem

Versions of this theorem are found in many undergraduate texts (Lifshitz and Pitaevskii 1981; Waldram 1985, p. 39), purporting to show that the Boltzmann entropy always increases. In the 'quantum' form of the theorem one writes the change of Boltzmann entropy in terms of the microstates α of the 1-particle system as

$$\frac{\mathrm{d}S_{\mathrm{B}}}{\mathrm{d}t} = -k_{\mathrm{B}}N \sum_{\alpha} \log p_1^{\alpha} \frac{\mathrm{d}p_1^{\alpha}}{\mathrm{d}t}. \tag{7.13}$$

In fact, in the original example quoted to me by the undergraduate at Cambridge, the errors in this theorem were compounded by calling S the Boltzmann–Gibbs entropy!

One is then invited to consider the 1-particle system(s) making 'quantum jumps' between the 1-particle microstates. The master equation and the principle of detailed balance for the transition rates $\nu_{\alpha\beta}$ then imply

$$\frac{\mathrm{d}S_{\mathrm{B}}}{\mathrm{d}t} = Nk_{\mathrm{B}} \sum_{\alpha\beta} \nu_{\alpha\beta}\big(\log p_{\beta} - \log p_{\alpha}\big)\big(p_{\beta} - p_{\alpha}\big) \geq 0. \tag{7.14}$$

What can one say about such a proof? There are several things wrong.

1. The use of approximate quantum mechanics, which is not necessarily valid for large perturbations.
2. Worse, it is *bad* quantum mechanics. An N-particle system has N-particle states. An isolated system will presumably sit in one of its N-particle microstates and make no transitions at all.
3. Even if you could prove such a theorem it would not be useful unless the change of entropy integrated over time were *numerically* equal to the change of experimental entropy. From the discussion of the last section it is clear that this cannot be the case.
4. Last, but not least, there are counter-examples to the theorem! The free expansion of molecular oxygen at 45 atmospheres and 160 K provides such an example, found by Jaynes (1971).

7.4.2 The second law of thermodynamics

The psychological need for an H-theorem is related to another misconception, one that concerns the second law of thermodynamics. For an isolated system the experimental entropy can only increase, that is

$$\Delta S_{\mathrm{E}} \geq 0, \tag{7.15}$$

with equality only if any changes are reversible.

The misconception this time is that, just because the experimental entropy has to increase, the theoretical entropy increases also. In fact, the Gibbs entropy S_{G} is actually a constant of the motion. This follows from Liouville's theorem for a classical system, or in the quantum case from the fact that the system will remain in an N-particle eigenstate. This dynamical constancy of the Gibbs entropy has sometimes been considered a weakness, but it is not. Remarkably, the constancy of the Gibbs theoretical entropy is exactly what one needs to *prove* the second law.

Once again, we return to the specific case of a gas of N particles, this time confined to one side of a box containing a removable partition. We suppose that the initial state is such that we can describe it using the canonical probability distribution. From our earlier discussion we can then say that the Gibbs entropy S_{G} is maximized and equal to the experimental entropy S_{E}.

We now suppose that the partition is opened and the atoms occupy the whole box. We wait until the state variables stop changing, so in that sense the system is in equilibrium and a new experimental entropy S'_{E} can be defined. Also, all the motions of the gas molecules are Hamiltonian, so that the Gibbs entropy S'_{G} has not changed: $S'_{\mathrm{G}} = S_{\mathrm{G}}$.

The probability distribution of the N particles is no longer the canonical one, however, because of the (very subtle!) correlations it contains reflecting the fact that the molecules were originally on one side of the partition. This means that the Gibbs entropy S'_{G} is now in general *less* than the maximum attainable for the new values of the state variables, which is in turn equal to the new experimental entropy. So

$$S_{\mathrm{E}} = S_{\mathrm{G}} = S'_{\mathrm{G}} \leq S'_{\mathrm{E}}. \tag{7.16}$$

This shows the fundamental result $S_{\mathrm{E}} \leq S'_{\mathrm{E}}$ and displays the second law of thermodynamics as a law concerning *experimental* quantities.

7.4.3 The theoretical second law

Another very important way of understanding the second law is to see it as a statement about phase volumes. Boltzmann's gravestone is engraved with the famous formula $S = k_{\mathrm{B}} \log W$. The W in this formula is the number of microstates compatible with the macroscopic state. This epitaph was placed there by Planck and it is ironic that this (correct) formula leads at once to Gibbs' definition of the entropy rather than Boltzmann's own.

Imagine, as in Fig. 7.1, the set of microstates compatible with the initial macroscopic state variables (p_1, T_1, etc.). This phase volume decribes our ability to reproduce the initial conditions: our system will be in a microstate somewhere inside this volume, but we do not know where. As the system evolves, the state variables change and finally reach new values (p_2, T_2, etc.). Our system has evolved dynamically and is now located

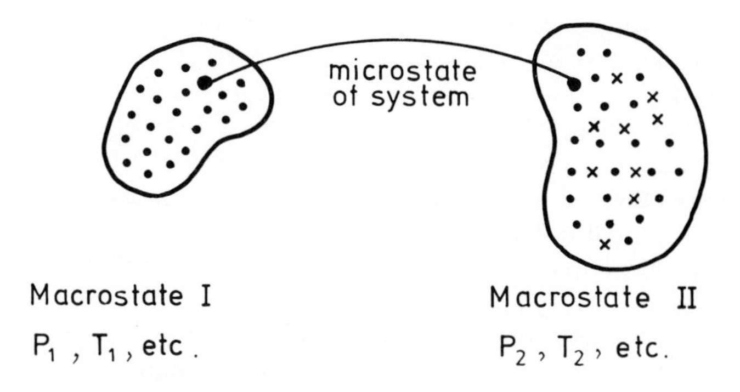

Fig. 7.1. Macrostates and microstates.

somewhere inside the phase volume consistent with these new values. This simple picture reveals a fundamental requirement for a process to be reproducible, as follows.

> The phase volume compatible with the final state cannot be less than the phase volume compatible with the initial state.

If the final phase volume were smaller there would be certain initial microstates that evolved to thermodynamic states other than (p_2, T_2); i.e., the thermodynamic process would not be reproducible. We have also the condition for reversibility:

> A process is reversible if the final phase volume is the same as the initial phase volume.

If the final phase volume were larger, then there would necessarily be some states in it that did not arise from states compatible with the initial state variables, and hence the process could not be reversible.

This completes our second, *theoretical* statement of the second law in terms of phase volumes.

7.5 Non-equilibrium phenomena

7.5.1 Time asymmetry in physics

There has been much debate about the nature of irreversibility and the 'arrow of time'. What has not been generally recognized is that temporal asymmetry enters because our knowledge of a system is not time-symmetric, and not because of any asymmetry inherent in its dynamics.

It should be stressed that I am not claiming that all physical laws must necessarily be time-symmetric, but merely that the ones we know of and need to consider here happen to have this property. Nevertheless, it is certainly true that our knowledge of the state of a system is not symmetric; we usually have more knowledge of its past state than of its future. This asymmetry in our knowledge is then properly reflected in asymmetry of our inferences about its likely behaviour. Once again, the problem is one of epistemology, not ontology.

Another related question concerns the Gibbs algorithm. It is recognized as a fine way of setting up an equilibrium ensemble, but how must it be modified to cope with disequilibria? The astonishing answer to this is also the simplest: the Gibbs algorithm is already complete; just give the formalism some time-dependent information and it will predict how the system is likely to behave and approach equilibrium.

7.5.2 Brownian motion

Rather than discuss generalities, which can be found elsewhere (Jaynes 1983; Grandy 1987; Garrett (Chapter 6 of this volume)), I can best illustrate the claims made above by a case-study, namely Brownian motion. Suppose a particle moves in one dimension, having position $x(t)$, and experiences random collisions from molecules at temperature T. We have previously considered a viewpoint that regards the microstate as a single point in an initial phase space, which thereafter moves deterministically as dictated by the Hamiltonian. For the present purposes we will abandon this Hamiltonian view and adopt a different approach which is able to cope with the outside influences. We consider instead the phase space to be described classically by $\mathrm{pr}\big[x(t)\big]$. Our knowledge of the particle's position is now encoded by this much larger joint probability distribution, which has a dimension for each moment of time. Our (incomplete) knowledge of the dynamics of the particle has to enter via constraints on the joint p.d.f.. As in path integral methods, we restrict our attention to the positions $x_n \equiv x(t_n)$ at a set of regularly spaced times $t_n = n\tau$. We define an average velocity $v_n \equiv \big(x_{n+1} - x_n\big)/\tau$ and acceleration $a_n \equiv \big(x_{n+1} - 2x_n + x_{n-1}\big)/\tau^2$. These definitions provide linear operators in x-space corresponding to velocity and acceleration. The slightly asymmetric definition of velocity is of no consequence in what follows, since the results are identical for the alternative definition $v_n \equiv \big(x_n - x_{n-1}\big)/\tau$.

We now use the Gibbs algorithm to set up an equilibrium ensemble by maximizing the entropy

$$S\big(\mathrm{pr}(\boldsymbol{x})\big) = -\int \mathrm{pr}(\boldsymbol{x}) \log \mathrm{pr}(\boldsymbol{x})\, \mathrm{d}\boldsymbol{x}. \tag{7.17}$$

In this definition we have dropped the dimensional factor of k_{B} and assumed a uniform measure over x-space.

We now introduce constraints suitable for the Brownian motion problem. Because the system is in equilibrium at temperature T we have, for all times t_n

$$\int \mathrm{d}\boldsymbol{x}\,\mathrm{pr}(\boldsymbol{x})v_n^2 \equiv \langle v_n^2 \rangle = \frac{k_{\mathrm{B}}T}{m}. \tag{7.18}$$

It is in fact only necessary, and certainly more convenient, to introduce a much weaker, single constraint, namely

$$\frac{1}{N_\tau}\sum_n \langle v_n^2 \rangle = \frac{k_{\mathrm{B}}T}{m}, \tag{7.19}$$

where N_τ is the number of time intervals considered.

We now introduce some knowledge of the dynamics. The colliding molecules can only provide a certain average impulse P to the particle in our time interval τ, so suppose in a similar way that

$$\frac{1}{N_\tau}\sum_n \langle a_n^2 \rangle = \left(\frac{P}{m\tau}\right)^2. \tag{7.20}$$

This specification of the average momentum transfer P certainly lies in the realm of dynamics, not inference. We suppose in what follows that it is sufficient to specify only one (P, τ) pair to describe all the reproducible features of Brownian motion. This may or may not be the case: only subsequent observation could tell us. If necessary, more information could be added in the form of further constraints. For example, the average impulse for time intervals of 2τ could be specified as well, in order to incorporate further details of the collision process.

I add one further constraint for the convenience of my computer program, which has the effect of confining the particle on average to a box of size L:

$$\frac{1}{N_\tau}\sum_n \langle x_n^2 \rangle = L^2. \tag{7.21}$$

Maximizing the entropy under these constraints yields a joint p.d.f. that represents the equilibrium ensemble,

$$\mathrm{pr}(\boldsymbol{x}) = \frac{1}{Z(\alpha,\beta,\gamma)}\exp\Big(-\boldsymbol{x}^{\mathrm{T}}(\alpha\mathbf{I} + \beta\mathbf{v}^{\mathrm{T}}\mathbf{v} + \gamma\mathbf{a}^{\mathrm{T}}\mathbf{a})\boldsymbol{x}/2\Big). \tag{7.22}$$

In the above, $\mathbf{v}$ and $\mathbf{a}$ are the velocity and acceleration operators (matrices) implicitly defined earlier, $\mathbf{I}$ is the identity matrix and α, β and γ are Lagrange multipliers. These multipliers control the physical variables of position, velocity and acceleration respectively: for example, β is the inverse temperature and γ provides the particle with inertia. The partition function

$$Z(\alpha,\beta,\gamma) = \int \mathrm{d}\boldsymbol{x} \, \exp\Big(-\boldsymbol{x}^{\mathrm{T}}(\alpha\mathbf{I} + \beta\mathbf{v}^{\mathrm{T}}\mathbf{v} + \gamma\mathbf{a}^{\mathrm{T}}\mathbf{a})\boldsymbol{x}/2\Big) \tag{7.23}$$

provides the normalizing integral and can be evaluated using a z-transform. The multipliers α, β and γ can be found by the usual partition function manipulations.

It is interesting to note that this joint p.d.f. for $\boldsymbol{x}$ is of the form $\exp\big(-\boldsymbol{x}^{\mathrm{T}}\mathbf{R}^{-1}\boldsymbol{x}/2\big)$, and can be recognized as a zero-mean, correlated, multivariate Gaussian time-series of a type well studied in digital signal processing: it is in fact an auto-regressive process of order 2. The covariance matrix of the time-series is given by $\big\langle \delta\boldsymbol{x}\delta\boldsymbol{x}^{\mathrm{T}}\big\rangle = \mathbf{R}$.

However, the equilibrium ensemble stands ready to deliver all sorts of time-dependent predictions: we just have to give it more data. Suppose, for example, that we know the position of the particle at various times: $x(t_1) = x_1 \pm \delta_1$ and $x(t_2) = x_2 \pm \delta_2$. We now employ our formalism to manipulate the probability distribution, given these data D:

$$\mathrm{pr}(\boldsymbol{x}, D) = \mathrm{pr}(D)\mathrm{pr}(\boldsymbol{x}|D) = \mathrm{pr}(\boldsymbol{x})\mathrm{pr}(D|\boldsymbol{x}). \tag{7.24}$$

In the above, $\mathrm{pr}(\boldsymbol{x}|D)$ is the answer we want, $\mathrm{pr}(\boldsymbol{x})$ is the equilibrium ensemble and $\mathrm{pr}(D|\boldsymbol{x})$ is the likelihood of the given data,

$$\mathrm{pr}(D|\boldsymbol{x}) \propto \exp\bigg(-\sum_{i=1,2} \frac{(x_i - X_i)^2}{2\delta_i^2}\bigg). \tag{7.25}$$

For the present purposes, $\mathrm{pr}(D)$ is an irrelevant normalizing constant. Bayes' theorem then gives us the answer $\mathrm{pr}(\boldsymbol{x}|D)$, showing the evolution of $\langle x(t)\rangle$ and $\big\langle(\delta x)^2(t)\big\rangle$ forwards and backwards in time.

In Fig. 7.2a, we have provided the information $x(84\tau) = 10 \pm 0.01$ to an ensemble with $\alpha \to 0$ and $\gamma \to 0$ (a system with no inertia). We know where the particle is at a certain time, so the display, which gives the average position and pointwise marginal uncertainty as a function of time, shows uncertainties which increase proportional to $t^{\frac{1}{2}}$ away from that time, in the manner of a random walk. The ensemble average position shows no net flux yet, because we only have this one piece of information. In particular, we do not yet know the velocity, although do we know that its magnitude is likely to have a value corresponding to thermal equilibrium.

In Fig. 7.2b we have added the extra information $x(42\tau) = 0 \pm 0.01$ to this same ensemble. The symmetrical random-walk behaviour persists for $t < 42\tau$ and $t > 84\tau$ but between these times, an average flux appears.

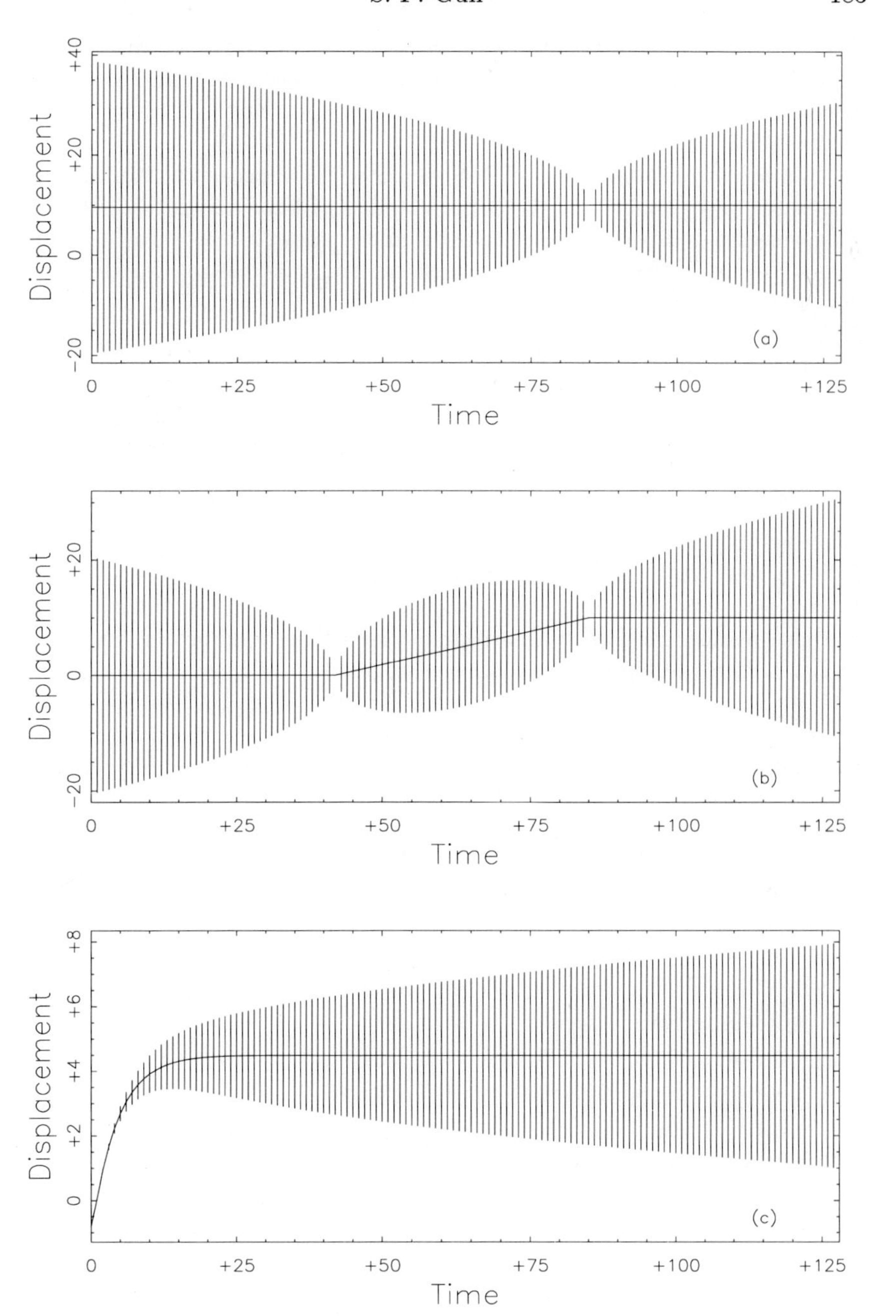

Fig. 7.2. Average positions and marginal uncertainties for a particle undergoing Brownian motion in one dimension.

The particle does, after all, have to go from $x = 0$ to $x = 10$ during this time interval and, in the absence of inertia, it is predicted to have travelled at constant velocity, though notice that our uncertainty about its position increases in the middle of this interval.

In our third example (Fig. 7.2*c*), we show the approach to equilibrium of a particle with inertia ($\gamma \neq 0$) projected with known velocity at $t = 0$. The average velocity decreases exponentially—indeed it satisfies a Langevin equation—whilst the uncertainty increases. This time the retrodictions are not plotted; the reader is invited to ponder what they look like, and what, if anything, they represent.

In these examples we have displayed the overall *marginal* uncertainty in the particle positions. The posterior p.d.f. is, however, highly correlated and there is important, additional information contained in these correlations. A very good way of visualizing this information is to plot typical samples from the posterior p.d.f. as well as the ensemble average.

7.6 Uncertainty versus fluctuations

Suppose we use the Gibbs algorithm to set up an equilibrium ensemble, and calculate the ensemble average of a quantity of interest f, together with its variance $(\Delta f)^2 \equiv \langle (f - \langle f \rangle)^2 \rangle$. Now Δf certainly represents our uncertainty about the quantity f but, according to most expositions of statistical mechanics, it is also supposed to indicate the level of temporal fluctuations of f. Here again, then, is a misconception—the fact that we are uncertain about the value of a quantity does not by itself mean that it must be fluctuating! Of course, it might be fluctuating and if that were the case, it would be a very good reason to be uncertain about its value. Without further analysis, however, we simply do not know whether it actually fluctuates. We have at last found a question in statistical mechanics where ergodic considerations are important. We can sketch a partial answer to this problem following Jaynes (1979).

We define

$$\bar{f} = \frac{1}{T} \int_0^T f(t)\, \mathrm{d}t \tag{7.26}$$

as a long-term time average and

$$(\delta f)^2 = \frac{1}{T} \int_0^T \left(f(t) - \bar{f} \right)^2 \mathrm{d}t \tag{7.27}$$

as a long-term variance. Taking ensemble averages, we do indeed find that $\langle f \rangle = \langle \bar{f} \rangle$; however

$$\langle (\delta f)^2 \rangle = (\Delta f)^2 + (\Delta \bar{f})^2, \tag{7.28}$$

and this second term is not necessarily zero.

The situation is as follows: if a time average is taken over too short a time interval, then the observed variation in f can of course be *less* than the Δf of the equilibrium ensemble. However, the long-term variation of f can actually be *greater* than Δf, depending on a particular property of the p.d.f. of the ensemble. Even then, although we can calculate $\langle \bar{f} \rangle$ and $\langle (\delta f)^2 \rangle$ as above, we still do not know that these estimates are reliable; to do that we have to examine higher-order correlations of the ensemble. The details are again in Jaynes (1979).

The moral is that the Gibbs algorithm gives the **uncertainty** of our predictions, not the observed temporal **fluctuation**. To say that a thermodynamic quantity actually fluctuates (which, of course, it may well do) requires further, decidedly non-trivial, analysis.

7.7 Conclusion

Most misconceptions about entropy result from misunderstandings about the role of probability theory and inference in physics. The epistemological content of statistical mechanics must be recognized and clearly separated from the ontological, dynamical aspects. Having done this, Gibbs' maximum entropy algorithm attains its full power, predicting the temporal evolution of *our state of knowledge* of a physical system.

Acknowledgements

I am extremely grateful to Geoff Daniell for helping me realize that I was a Bayesian and thereby introducing me to the work of Ed Jaynes.

References

Cox, R. T. (1946). Probability, frequency and reasonable expectation. *American Journal of Physics*, **14**, 1–13.

Grandy, W. T. (1987). *Foundations of statistical mechanics*, Vols 1 and 2. Reidel, Dordrecht.

Gull, S. F. and Skilling, J. (1984). The maximum entropy method in image processing. *IEE Proceedings*, **131F**, 646–59.

Jaynes, E. T. (1965). Gibbs vs Boltzmann entropies. *American Journal of Physics*, **33**, 391–8.

Jaynes, E. T. (1971). Violation of Boltzmann's H theorem in real gases. *Physical Review A*, **4**, 747–50.

Jaynes, E. T. (1979). Where do we stand on maximum entropy? In *The maximum entropy formalism* (ed. R. D. Levine and M. Tribus), pp. 15–118. M.I.T. Press, Cambridge, Massachusetts.

Jaynes, E. T. (1982). On the rationale of maximum entropy methods. *Proceedings of the IEEE*, **70**, 939–52.

Jaynes, E. T. (1983). *Papers on probability, statistics and statistical physics*, Synthese Library, Vol. 158 (ed. R. D. Rosenkrantz). Reidel, Dordrecht.

Lifshitz, E. M. and Pitaevskii, L. P. (1981). *Physical kinetics*. Pergamon, Oxford.

Shore, J. E. and Johnson, R. W. (1980). Axiomatic derivation of the principle of maximum entropy and the principle of minimum cross-entropy. *IEEE Transactions on Information Theory*, **IT-26**, 26–37.

Waldram, J. R. (1985). *The theory of thermodynamics*. Cambridge University Press.

8

The X-ray crystallographic phase problem

G. Bricogne

Abstract

The X-ray analysis of crystal structure may be viewed as an inverse problem in which the three-dimensional structure of a crystal is to be reconstructed from the results of an X-ray diffraction experiment. The intensities of the scattered beams which can be measured in such an experiment are essentially the squared moduli of the Fourier coefficients associated with the triply periodic electron density distribution in the crystal. Unfortunately the corresponding phases cannot be measured, yet they must be restored by some means before the desired picture of the crystal structure can be obtained by Fourier synthesis. This constitutes the 'phase problem' of X-ray crystallography.

Because of its extreme practical importance, the phase problem has been under constant attack for over three quarters of a century, but there is still no generally applicable method for solving it. Its recalcitrant nature derives from the fact that it is mathematically indeterminate unless sufficient *chemical* information is brought to bear on its solution to constrain it to a single reasonable answer, while this chemical information cannot be adequately handled by any of the well-developed devices of mathematics until a stereochemical parametrization of the electron density becomes possible.

A statistical approach to the phase problem, initiated in the early fifties and later developed into successful direct methods for the determination of small structures, has recently been improved and extended by invoking the powerful 'saddlepoint method' of analytic probability theory. An essential role is played by the maximum entropy property of certain probability distributions. This role, however, is limited to that of a useful computational device, and the maximum entropy property itself appears only as a by-product of the saddlepoint method rather than as a principle of cosmic significance.

8.1 Introduction

Early crystal structure determinations relied extensively on *a priori* chemical knowledge in order to turn the experimental diffraction measurements into a chemical model. Only relatively late in the progress of X-ray analysis was it realized that some general, non-chemical properties of the electron density in a crystal—namely *positivity* (Harker and Kasper 1948; Karle and Hauptman 1950) and *atomicity* (Sayre 1952)—imply mathematical relationships between the phases and the amplitudes of the Fourier coefficients, so that a direct mathematical attack on the phase problem could be envisaged.

The most fruitful of these lines of attack turned out to reside in a *statistical* formulation of the phase problem, in which the property of atomicity is abstracted to the statement that crystal structures may be conceived as being generated by placing atoms randomly, and independently of each other, throughout the unit cell of the crystal. The Fourier coefficients pertaining to such structures are therefore sums of a large number of random variables (the contributions from individual atoms), and the limit theorems of probability theory may then be invoked to yield estimates of their distribution.

These ideas were first put to use by Wilson (1949) in his study of the distribution of diffracted intensities and of its modulation by crystal symmetry. The next logical step was taken by Hauptman and Karle (1953), who examined the *joint* distribution of several Fourier coefficients, and made the crucial observation that such distributions could yield phase information if the amplitudes were assumed to be known. Bertaut (1955*a,b,c*) and Klug (1958) later reexamined their initial formulation, and showed that its mathematical foundations were to be found in the well-developed theory of normal approximation by means of the Gram–Charlier or Edgeworth asymptotic expansions (Cramér 1946).

Direct methods underwent a remarkable development between the mid-sixties and the mid-seventies with the advent of the symbolic addition method (Karle and Karle 1966) and of multisolution strategies (Germain and Woolfson 1968) for the systematic exploitation of the 'triple-phase relationship'. These advances ultimately resulted in computer programs capable of solving routinely most organic structures containing up to 100 atoms. During the next decade or so, steady progress was made in the derivation and exploitation of more elaborate formulae. In spite of these improvements, the solution by direct methods of structures containing over 150 atoms remained the exception rather than the rule. It seemed difficult, however, to ascertain whether this upper bound represented a fundamental limit to the intrinsic power of the statistical approach, or was merely a consequence of some temporary shortcomings in its conventional implementation.

This question was settled in favour of the second alternative (Bricogne 1984) through a reappraisal of the mathematical foundations of probabilistic direct methods. This study indicated the possibility of a major increase in their power if a new analytical tool (the so-called 'saddlepoint method') was adopted to replace the previous ones. At the same time, it brought to light an intimate connection between this enhancement of direct methods and a mode of statistical inference based on Jaynes' maximum entropy principle (Jaynes 1957, 1968).

In this paper I will first present a quick informal overview (Section 8.2) of the main ideas which underlie these 'direct methods' of phase determination, including recent developments related to the maximum entropy method. I will then focus on the fundamental mathematical construct—the joint probability distribution of structure factors—on which the entire approach is based. To make the presentation self-contained and to establish both notation and terminology, I will review (Section 8.3) some of the basic analytical methods of probability theory. I will then show (Sections 8.4–8.6) how the application of these methods to the phase problem leads naturally to the granting of a privileged status to prior distributions of random atomic positions which have maximum entropy under linear constraints reflecting various phase assumptions. This status, however, is limited to that of a useful computational device: the maximum entropy property appears only as a by-product of the saddlepoint method, and its use can be justified on purely analytical grounds without invoking the customary mixture of philosophy and heuristics subsumed under the name of the 'maximum entropy principle'.

8.2 Overview

8.2.1 Preamble

The maximum entropy method is not a completely different approach from conventional direct methods: it is an *improvement* of the latter at the level of the techniques of analytic probability theory on which they are based.

Its first attraction is that it uses almost exclusively the 'high-level language' of classical probability theory and statistics with a minimum of *ad hoc* jargon, and relegates to a subsidiary role the algebraic *minutiae* of phase invariants etc., which make direct methods seem so very esoteric to outsiders (see, for example, Hauptman (1972)): if the maximum entropy approach proves successful, the earlier methods should ultimately be 'phased out'. Its main attraction, however, is that it is in principle much more powerful than conventional direct methods (although a lot of work still has to be done to turn this potential into a tangible reality), especially in the macromolecular field.

One of the leading figures in the development of conventional direct methods, Prof. Michael Woolfson of York University, wrote (Woolfson 1987):

> 'Entropy maximization is adding nothing new to the crystallographic scene, and, since it involves a great deal of effort, perhaps nothing useful.'

In the present author's view, this judgement constitutes the most compelling reason for believing that the maximum entropy method will rapidly supersede direct methods as we know them today.

8.2.2 The phase problem

Let $\rho(\boldsymbol{x})$ denote the electron density at position $\boldsymbol{x}$ in the unit cell of the crystal, where $\boldsymbol{x} = (x_1, x_2, x_3)$ is a triple of crystallographic (or *fractional*) coordinates and ρ is expressed in the dimensionless unit of electrons per unit cell. By the definition of a crystal and of fractional coordinates, ρ is periodic with period lattice the integer lattice $\mathbb{Z}^3$, and hence can be written as a triple Fourier series, with (usually complex) coefficients $F_{\mathbf{h}}$ called the *structure factors*:

$$\rho(\boldsymbol{x}) = \sum_{\mathbf{h} \in \mathbb{Z}^3} F_{\mathbf{h}} \exp(-2\pi i \mathbf{h} \cdot \boldsymbol{x}), \tag{8.1}$$

$$F_{\mathbf{h}} = \int_{[0,1]^3} \rho(\boldsymbol{x}) \exp(2\pi i \mathbf{h} \cdot \boldsymbol{x})\, d^3\boldsymbol{x}. \tag{8.2}$$

The kinematic approximation to the theory of X-ray diffraction shows that the intensity $I_{\mathbf{h}}$ of a reflection labelled by Miller indices $\mathbf{h}$ is proportional to $|F_{\mathbf{h}}|^2$ and does not depend on the phase $\phi_{\mathbf{h}}$ of $F_{\mathbf{h}}$; the latter must therefore be obtained by some other means before $\rho(\boldsymbol{x})$ can be computed from (8.1), and the structure visualized.

The phase problem *per se*, that is, posed abstractly as the problem of computing phases from moduli data alone, is mathematically indeterminate. However, not all combinations of phases are equally acceptable when solving a crystal structure: a chemist can tell the difference between good phases (which produce an electron density map $\rho(\boldsymbol{x})$ with atom-like features connected in a stereochemically sensible way) and bad phases (for which the map does not make chemical sense).

This criterion of *chemical validity* is very hard to translate into a tractable mathematical property of the structure factors. Therefore we will have to *weaken* it

- *enough*, for the mathematics to become tractable;

- *not too much*, in order to retain some power of discrimination.

Clearly, any method obtained in this way can only act as a *sieve*, since it will still accept some non-chemical solutions: an extra step of interpretation and validation will usually be needed.

8.2.3 Atomicity and the random atom model

A felicitous compromise between power and tractability is provided by the property of *atomicity*. This word has two distinct meanings in conventional direct methods:

- as used by Sayre (1952), it results in an *exact analytical* relation between the structure factors of *every* structure;
- as used here, it gives rise to *statistical properties* which are obeyed only when viewed over an *ensemble* of structures (hence the forthcoming connection with entropy).

Surprisingly, the assumption that a crystal structure is made up of atoms gives rise by itself to statistical constraints on the F's, *even if we do not know where the atoms are.*

8.2.4 The basic idea

Hauptman and Karle (1953) gave the first formulation of a procedure for the direct determination of phases, solely from moduli data. Although new to crystallographers, their methods bore a close resemblance to those developed with great vigour during the previous decade for designing radar detection systems, which laid the foundations for the statistical theory of communication (see, for example, Rice (1944, 1945)). The procedure goes as follows:

1. Derive $\mathcal{P}(F_{\mathbf{h}_1}, \ldots, F_{\mathbf{h}_n})$, the joint probability distribution (j.p.d.) of the structure factors, denoted $\mathcal{P}(\boldsymbol{F})$ in the sequel, from the random atom model.
2. Put $|F_{\mathbf{h}_j}| = |F_{\mathbf{h}_j}|^{\mathrm{obs}}$ in each argument of $\mathcal{P}$.
3. Get $\mathrm{Prob}(\phi_{\mathbf{h}_1}, \ldots, \phi_{\mathbf{h}_n} \mid |F_{\mathbf{h}_1}|, \ldots, |F_{\mathbf{h}_n}|)$, the conditional probability distribution (c.p.d.) of the phases given the observed values of the moduli, as (up to normalization)

$$\mathcal{P}(|F_{\mathbf{h}_1}| \exp(\mathrm{i}\phi_{\mathbf{h}_1}), \ldots, |F_{\mathbf{h}_n}| \exp(\mathrm{i}\phi_{\mathbf{h}_n})).$$

Thus, *given the observed moduli*, we get indications that *certain combinations of phases are more probable than others*, which is just what we want. This seemingly straightforward idea is however unexpectedly difficult to implement in an optimal fashion.

8.2.5 The basic shortcoming

Conventional direct methods obtain $\mathcal{P}(\boldsymbol{F})$ for uniformly distributed random atoms (by methods to be examined later) as a *series expansion* which is most accurate for *small deviations* from its mean value $\langle \boldsymbol{F} \rangle = \mathbf{0}$, that is, for small moduli only; and yet these distributions are always used after substituting the *largest* moduli available. The problem cannot be cured by summing more terms, because the series in question is an *asymptotic series.*

To make matters worse, conventional direct methods have never made use of $\mathcal{P}(\boldsymbol{F})$ as a whole, but have instead considered *individual terms* of the series, and then have spent a lot of effort trying to put the series back together again by questionable methods—with problems of non-independence (see Bricogne (1984, Section 2.2)).

8.2.6 Consequences

The analysis just given has two consequences:

1. There is *no such thing*, for practical purposes, as a unique functional expression for the j.p.d. $\mathcal{P}(\boldsymbol{F})$ (which had been naïvely assumed) into which observed moduli can be simply *substituted* to give the c.p.d. of phases.
2. We need an approximation method *other* than that used in conventional direct methods (called the Edgeworth series), since it must be able to work for *large deviations* from the expectation $\langle \boldsymbol{F} \rangle$ of $\boldsymbol{F}$.

8.2.7 The maximum entropy method

Two essential departures from conventional direct methods will characterize the maximum entropy approach which will enable it to deal with the problem of large deviations:

1. Different approximations to $\mathcal{P}$ will be used in different regions of $\boldsymbol{F}$-space, that is, for different phase assumptions made on large moduli. These are most conveniently indexed by a *phasing tree* which describes the hierarchical dependence between all the phase assumptions made.
2. Each approximation labelled by a node of the phasing tree will be calculated using the *saddlepoint method* rather than the Edgeworth series. It is at this stage that the notion of *entropy* is encountered, and that it is required to *maximize* it.

Therefore the buzz-word **maximum entropy** should not make us lose sight of the fact that what we are building is *an improvement of conventional direct methods in their own terms*: there are no new magic principles, no new justifications required; just more powerful analytical methods.

8.2.8 A tree-directed multisolution strategy

The precise object attached to each node ν of the phasing tree consists of:

- a *basis set* $H_\nu = \{\mathbf{h}_1^\nu, \ldots, \mathbf{h}_{n_\nu}^\nu\}$ of strong reflections;
- a set of *trial phase values* $\Phi_\nu = \{\phi_1^\nu, \ldots, \phi_{n_\nu}^\nu\}$ for them;
- an *approximation* to $\mathcal{P}(\boldsymbol{F})$ near $\boldsymbol{F} = \boldsymbol{F}_{H_\nu}^*$ allowing the accurate computation of conditional probabilities $\mathcal{P}(\boldsymbol{F}_K \mid \boldsymbol{F}_{H_\nu} = \boldsymbol{F}_{H_\nu}^*)$, where K is a set of unique non-origin reflections disjoint from H.

The use of a tree, a structure borrowed from AI (see, for example, Nilsson (1971)), invites an analogy with *game-playing programs* which, although it can be dangerously distracting, has the merit of drawing attention to the need for *two types of criteria* to assist phase determination:

- one to measure the strength of the current position: this will turn out to be *entropy*;
- one to measure the ability of the current position to lead to strong positions after more moves have occurred, that is, to provide a 'look-ahead' capability: this will turn out to be *likelihood*.

8.2.9 Calculation of joint probabilities

It is convenient to switch to unitary structure factors, which for equal point atoms of unit weight are given by $\boldsymbol{U}_\mathbf{h} = \boldsymbol{F}_\mathbf{h}/F_\mathbf{0}$ (see (8.2)).

The *saddlepoint approximation* $\mathcal{P}^{\mathrm{SP}}(\boldsymbol{U}_H^*)$ to $\mathcal{P}(\boldsymbol{U}_H^*)$ for N such atoms distributed in the unit cell with probability density $m(\boldsymbol{x})$ is given by

$$\mathcal{P}^{\mathrm{SP}}(\boldsymbol{U}_H^*) = \frac{\exp(N\mathcal{S})}{\left(\det(2\pi N\mathbf{Q})\right)^{\frac{1}{2}}}, \tag{8.3}$$

where

$$\mathcal{S} = \mathcal{S}_m(q^{\mathrm{ME}}). \tag{8.4}$$

Here

$$\mathcal{S}_m(q) = -\int q(\boldsymbol{x}) \log\left(\frac{q(\boldsymbol{x})}{m(\boldsymbol{x})}\right) \mathrm{d}^3\boldsymbol{x} \tag{8.5}$$

denotes the relative entropy of a probability distribution $q(\boldsymbol{x})$ with respect to $m(\boldsymbol{x})$, and q^{ME} maximizes $\mathcal{S}_m(q)$ under the constraints that its Fourier coefficients for the basis-set reflections should have the value $\boldsymbol{U}_H^*$. The matrix $\mathbf{Q}$ can be computed from the full set of Fourier coefficients of q^{ME} by structure factor algebra. The method of calculation of q^{ME} for a given $\boldsymbol{U}_H^*$ will be described later.

Note that this expression can be evaluated numerically *even if H contains hundreds, thousands, ..., hundreds of thousands of reflections.* This would be totally out of the question with the Edgeworth series.

8.2.10 Maximum entropy extrapolation

The process of forming the maximum entropy probability distribution q^{ME} from the basis-set data $\boldsymbol{U}_H^*$ enables *extrapolation* to Fourier coefficients $\boldsymbol{U}_{\mathbf{k}}^{\mathrm{ME}}$ for $\mathbf{k} \notin H$. This extrapolation allows *phase extension* (see Bricogne (1984, Section 7.3)) in circumstances where conventional direct methods—including tangent formulae—give nothing, or diverge.

However, starting from poor basis-set phases would produce worse, and probably useless, extrapolated phases: we need to be able to *refine* phases before taking their extrapolates seriously. This is where we need the look-ahead capability provided by *likelihood.*

The idea is to try and consult the unphased moduli outside the basis set in order to select or refine the basis set phases we have chosen in $\boldsymbol{U}_H^*$. Since this is also the purpose for which Hauptman's *neighbourhoods* (Hauptman 1980) and Giacovazzo's *phasing shells* (Giacovazzo 1980) were defined, it will be no surprise to find that they are somehow present implicitly in this likelihood criterion.

8.2.11 Calculation of conditional probabilities

Let K be a set of non-basis reflections ($H \cap K = \emptyset$) and let $\boldsymbol{U}_K$ be the corresponding vector of structure factors. Then the conditional distribution of $\boldsymbol{U}_K$ given that $\boldsymbol{U}_H = \boldsymbol{U}_H^*$ is

$$\mathcal{P}^{\mathrm{SP}}\left(\boldsymbol{U}_K \mid \boldsymbol{U}_H = \boldsymbol{U}_H^*\right) \propto \exp\left(-\tfrac{1}{2}N\left(\boldsymbol{U}_K - \boldsymbol{U}_K^{\mathrm{ME}}\right)^{\mathrm{T}}\mathbf{Q}_{KK}^{-1}\left(\boldsymbol{U}_K - \boldsymbol{U}_K^{\mathrm{ME}}\right)\right), \tag{8.6}$$

that is, it is a multivariate Gaussian centred at $\boldsymbol{U}_K^{\mathrm{ME}}$ (not at $\mathbf{0}_K$) whose covariance matrix $\mathbf{Q}_{KK}$ can be computed from the Fourier coefficients of q^{ME} (including the extrapolates) by structure factor algebra.

This conditional distribution is in general strongly *multimodal*, that is, it has many local maxima, making the process of phase extension intrinsically ambiguous ('branching problem'). Fortunately, the phasing tree constitutes an ideal book-keeping device to handle this ambiguity too. The mechanism by which these constantly updated distributions confer an advantage over conventional direct methods is identical to the basis of the 'card-counting' strategy at the game of blackjack (Thorpe 1966): it resides in the optimal use, at each stage, of all the information available. In the crystallographic context, the current estimate of the non-uniformity of q^{ME} plays the same role as the knowledge of the current composition of the pack of cards in enabling the 'player' to make the most unbiased assessments of uncertainty.

The probability distribution one would derive from the null hypothesis $\boldsymbol{U}_H = \mathbf{0}$ would be the usual Gaussian centred at the origin:

$$\mathcal{P}(\boldsymbol{U}_K \mid \boldsymbol{U}_H = \mathbf{0}) \propto \exp\left(-\tfrac{1}{2} N \boldsymbol{U}_K^{\mathrm{T}} \boldsymbol{\epsilon}_{KK}^{-1} \boldsymbol{U}_K\right), \tag{8.7}$$

where $\boldsymbol{\epsilon}_{KK}$ is the diagonal matrix of statistical weights (the Wilson distribution).

8.2.12 Likelihood and likelihood ratio

The two distributions just derived under two different hypotheses on $\boldsymbol{U}_H$ cannot be easily compared because they involve the phases $\phi_{\mathbf{k}}$ of non-basis reflections. However, we can *integrate these phases out* to obtain the *conditional probability distributions* not of the structure factors $\boldsymbol{U}_K$ but *of their moduli* $|\boldsymbol{U}_K|$.

We may then compare the conditional probabilities assigned to the *observed* values of moduli under each of the two hypotheses, the better assumption being that which assigns the *highest* probability to them. We may use this quantity, called the *likelihood* of the hypothesis, as a figure of merit.

The comparison is best carried out using the likelihood ratio

$$\frac{\Lambda(\boldsymbol{U}_H^*)}{\Lambda(\mathbf{0})} = \frac{\mathcal{P}(|\boldsymbol{U}_K|^{\mathrm{obs}} \mid \boldsymbol{U}_H = \boldsymbol{U}_H^*)}{\mathcal{P}(|\boldsymbol{U}_K|^{\mathrm{obs}} \mid \boldsymbol{U}_H = \mathbf{0})}, \tag{8.8}$$

or its logarithm, the *log-likelihood gain.* This is our 'look-ahead criterion' which plays the role of the heuristic function in AI. We could also invoke Bayes' theorem and calculate the *a posteriori* probability of $\boldsymbol{U}_H^*$ by

$$\mathcal{P}^{\mathrm{post}}(\boldsymbol{U}_H^*) \propto \mathcal{P}^{\mathrm{SP}}(\boldsymbol{U}_H^*) \times \Lambda(\boldsymbol{U}_H^*) \tag{8.9}$$

and *refine* the basis-set phases in $\boldsymbol{U}_H^*$ by maximizing this quantity with respect to them.

8.2.13 Comparison with conventional direct methods

The likelihood contains in an implicit form the entire theory of cross-terms, of neighbourhoods (Hauptman 1980) and of representations (Giacovazzo 1980). The definition of the neighbourhoods of the basis set is very natural (no need for a 'neighbourhood principle'): they consist of those non-basis reflections where the maximum entropy extrapolation is the strongest, and they can be ranked according to the power of $N^{-\frac{1}{2}}$ to which that mean strength is proportional. Note that we define in this way the neighbourhoods of the entire basis set, not only those of individual invariants; the latter are generated *numerically* in the course of entropy maximization and of integration over the non-basis phases, and never have to be handled explicitly.

As a simple example, the Hauptman quartet formula (Hauptman 1975) can be derived in a few lines for small U values, and can be made intuitively obvious (Bricogne 1984, Section 4.2.2; Bricogne 1988*b*, pp. 70–72). For large U values, however, only the likelihood and posterior probability formulations are quantitatively correct.

According to a fundamental theorem of Neyman and Pearson (1933), likelihood is a very special figure of merit as an aid to making optimal decisions. In the context of direct methods this implies that, although many quantities can be concocted which will be *useful* for selecting good phase sets and refining them (the York school seems to be an inexhaustible source of these), *likelihood is the best possible.* Thus the optimality of likelihood is not an advertising slogan, it is a classical theorem of statistics.

8.3 Analytical methods of probability theory

We will now focus our attention on the analytical devices used in formulating and implementing direct methods of phase determination. This section is not intended as an introduction to probability theory—for which the reader is referred to the books by Cramér (1946), Lindley (1965) or Bhattacharya and Ranga Rao (1976).

8.3.1 Convolution of probability densities

The addition of independent random variables or vectors leads to the convolution of their probability distributions: if $\boldsymbol{X}_1$ and $\boldsymbol{X}_2$ are two n-dimensional random vectors over the reals, independently distributed with probability densities P_1 and P_2 respectively, then their sum $\boldsymbol{X} = \boldsymbol{X}_1 + \boldsymbol{X}_2$ has probability density $\mathcal{P}$ given by

$$\mathcal{P}(\boldsymbol{X}) = \int_{\mathbb{R}^n} P_1(\boldsymbol{X}_1) P_2(\boldsymbol{X} - \boldsymbol{X}_1)\, \mathrm{d}^n \boldsymbol{X}_1 = \int_{\mathbb{R}^n} P_1(\boldsymbol{X} - \boldsymbol{X}_2) P_2(\boldsymbol{X}_2)\, \mathrm{d}^n \boldsymbol{X}_2,$$

that is,

$$\mathcal{P} = P_1 * P_2. \tag{8.10}$$

This result can be extended, by means of the theory of distributions, to the case where P_1 and P_2 are singular measures and do not have a density with respect to the Lebesgue measure in $\mathbb{R}^n$.

8.3.2 Characteristic functions

This convolution can be turned into a simple multiplication by considering the Fourier transforms, called the *characteristic functions*, of P_1, P_2 and $\mathcal{P}$. The Fourier transformation used here is defined with a slightly different normalization from the usual one in crystallography in that there is no factor of 2π in the exponent:

$$C(\boldsymbol{t}) = \int_{\mathbb{R}^n} P(\boldsymbol{X}) \exp(\mathrm{i}\boldsymbol{t} \cdot \boldsymbol{X})\, \mathrm{d}^n\boldsymbol{X}. \tag{8.11}$$

Then by the convolution theorem

$$\mathcal{C}(\boldsymbol{t}) = C_1(\boldsymbol{t}) \times C_2(\boldsymbol{t}), \tag{8.12}$$

so that $\mathcal{P}(\boldsymbol{X})$ may be evaluated by Fourier inversion of its characteristic function as:

$$\mathcal{P}(\boldsymbol{X}) = \frac{1}{(2\pi)^n} \int_{\mathbb{R}^n} C_1(\boldsymbol{t}) C_2(\boldsymbol{t}) \exp(-\mathrm{i}\boldsymbol{t} \cdot \boldsymbol{X})\, \mathrm{d}^n\boldsymbol{t}. \tag{8.13}$$

(Note the normalization factor.)

It follows from the differentiation theorem that the partial derivatives of the characteristic function $C(\boldsymbol{t})$ at $\boldsymbol{t} = \boldsymbol{0}$ are related to the moments of a distribution P by the identities

$$\mu_{r_1, r_2, \ldots, r_n} \equiv \int_{\mathbb{R}^n} P(\boldsymbol{X}) X_1^{r_1} X_2^{r_2} \cdots X_n^{r_n}\, \mathrm{d}^n\boldsymbol{X} \tag{8.14}$$

$$= \mathrm{i}^{-n} \left. \frac{\partial^{r_1 + \cdots + r_n} C}{\partial t_1^{r_1} \cdots \partial t_n^{r_n}} \right|_{\boldsymbol{t}=\boldsymbol{0}} \tag{8.15}$$

for any n-tuple of non-negative integers $(r_1, r_2, \ldots, r_n)$.

8.3.3 Moment-generating functions

The above relation can be freed from powers of i by defining (at least formally) the *moment-generating function*

$$M(\boldsymbol{t}) = \int_{\mathbb{R}^n} P(\boldsymbol{X}) \exp(\boldsymbol{t} \cdot \boldsymbol{X})\, \mathrm{d}^n\boldsymbol{X}, \tag{8.16}$$

which is related to $C(\boldsymbol{t})$ by $C(\boldsymbol{t}) = M(\mathrm{i}\boldsymbol{t})$ so that the inversion formula reads

$$\mathcal{P}(\boldsymbol{X}) = \frac{1}{(2\pi)^n} \int_{\mathbb{R}^n} M_1(\mathrm{i}\boldsymbol{t}) M_2(\mathrm{i}\boldsymbol{t}) \exp(-\mathrm{i}\boldsymbol{t} \cdot \boldsymbol{X})\, \mathrm{d}^n\boldsymbol{t}. \tag{8.17}$$

The moment-generating function is well defined, in particular, for any probability distribution with compact support, in which case it may be continued analytically from a function over $\mathbb{R}^n$ into an entire function of n *complex* variables by virtue of the Paley–Wiener theorem (Paley and Wiener 1934) (see also Bricogne (1984, Section 5.1)). Its moment-generating properties are summed up in the following relations:

$$\mu_{r_1,r_2,\ldots,r_n} = \left. \frac{\partial^{r_1+\cdots+r_n} M}{\partial t_1^{r_1} \cdots \partial t_n^{r_n}} \right|_{\boldsymbol{t}=\boldsymbol{0}}. \tag{8.18}$$

8.3.4 Cumulant-generating functions

The multiplication of moment-generating functions may be further simplified into the addition of their logarithms

$$\log \mathcal{M} = \log M_1 + \log M_2, \tag{8.19}$$

or equivalently of the coefficients of their Taylor series at $\boldsymbol{t} = \boldsymbol{0}$, viz.,

$$\kappa_{r_1,r_2,\ldots,r_n} = \left. \frac{\partial^{r_1+\cdots+r_n} (\log M)}{\partial t_1^{r_1} \cdots \partial t_n^{r_n}} \right|_{\boldsymbol{t}=\boldsymbol{0}}. \tag{8.20}$$

These coefficients are called *cumulants*, since they add when the independent random vectors to which they belong are added, and $\log M$ is called the *cumulant-generating function*. The inversion formula for $\mathcal{P}$ then reads

$$\mathcal{P}(\boldsymbol{X}) = \frac{1}{(2\pi)^n} \int_{\mathbb{R}^n} \exp\big(\log M_1(\mathrm{i}\boldsymbol{t}) + \log M_2(\mathrm{i}\boldsymbol{t}) - \mathrm{i}\boldsymbol{t} \cdot \boldsymbol{X}\big)\, \mathrm{d}^n\boldsymbol{t}. \tag{8.21}$$

8.3.5 Asymptotic expansions and limit theorems

Consider an n-dimensional random vector $\boldsymbol{X}$ of the form

$$\boldsymbol{X} = \boldsymbol{X}_1 + \boldsymbol{X}_2 + \cdots + \boldsymbol{X}_N, \tag{8.22}$$

where the N summands are independent n-dimensional random vectors identically distributed with probability density P. Then the distribution $\mathcal{P}$ of $\boldsymbol{X}$ may be written in closed form as a Fourier transform:

$$\begin{aligned} \mathcal{P}(\boldsymbol{X}) &= \frac{1}{(2\pi)^n} \int_{\mathbb{R}^n} M^N(\mathrm{i}\boldsymbol{t}) \exp(-\mathrm{i}\boldsymbol{t} \cdot \boldsymbol{X})\, \mathrm{d}^n\boldsymbol{t} & (8.23) \\ &= \frac{1}{(2\pi)^n} \int_{\mathbb{R}^n} \exp\big(N \log M(\mathrm{i}\boldsymbol{t}) - \mathrm{i}\boldsymbol{t} \cdot \boldsymbol{X}\big)\, \mathrm{d}^n\boldsymbol{t}, & (8.24) \end{aligned}$$

where

$$M(\boldsymbol{t}) = \int_{\mathbb{R}^n} P(\boldsymbol{Y}) \exp(\boldsymbol{t} \cdot \boldsymbol{Y})\, \mathrm{d}^n \boldsymbol{Y} \tag{8.25}$$

is the moment-generating function common to all the summands.

This is an exact expression for $\mathcal{P}$, but in most cases some form of approximation must be used to evaluate the integral. For this purpose it is customary (Cramér 1946; Bhattacharya and Ranga Rao 1976) to expand the cumulant-generating function around $\boldsymbol{t} = \mathbf{0}$ with respect to the carrying variables $\boldsymbol{t}$:

$$\log\bigl(M^N(\mathrm{i}\boldsymbol{t})\bigr) = \sum_{\mathbf{r}\in\mathbb{N}^n} \frac{N\kappa_{\mathbf{r}}}{\mathbf{r}!} (\mathrm{i}\boldsymbol{t})^{\mathbf{r}}, \tag{8.26}$$

where $\mathbf{r} = (r_1, r_2, \ldots, r_n)$ is an n-tuple of non-negative integers; $\mathbf{r}! = r_1!\, r_2! \cdots r_n!$; $\boldsymbol{t}^{\mathbf{r}} = t_1^{r_1} \cdots t_n^{r_n}$; and for later use $|\mathbf{r}| = r_1 + r_2 + \cdots + r_n$. The first-order terms may be eliminated by recentring $\mathcal{P}$ around its vector of first-order cumulants

$$\langle \boldsymbol{X} \rangle = \sum_{j=1}^{N} \langle \boldsymbol{X}_j \rangle, \tag{8.27}$$

where $\langle \cdots \rangle$ denotes the mathematical expectation of a random vector. The second-order terms may be grouped separately from the terms of third or higher order to give

$$M^N(\mathrm{i}\boldsymbol{t}) = \exp\bigl(-\tfrac{1}{2} N \boldsymbol{t}^{\mathrm{T}} \mathbf{Q} \boldsymbol{t}\bigr) \times \exp\Biggl(\sum_{|\mathbf{r}|\geq 3} \frac{N\kappa_{\mathbf{r}}}{\mathbf{r}!} (\mathrm{i}\boldsymbol{t})^{\mathbf{r}} \Biggr), \tag{8.28}$$

where $\mathbf{Q} = \nabla\nabla^{\mathrm{T}}(\log M)$ is the covariance matrix of the multivariate distribution $\mathcal{P}$. Expanding the second exponential gives rise to a series of terms of the form

$$\exp\bigl(-\tfrac{1}{2} N \boldsymbol{t}^{\mathrm{T}} \mathbf{Q} \boldsymbol{t}\bigr) \times \text{monomial in } t_1,\, t_2,\, \ldots,\, t_n. \tag{8.29}$$

Each of these terms may now be subjected to Fourier inversion, giving a product of Hermite functions of $\boldsymbol{t}$ with coefficients involving the cumulants κ of $\mathcal{P}$. Taking the transformed terms in natural order gives an asymptotic expansion of $\mathcal{P}$ for large N called the *Gram–Charlier series* of $\mathcal{P}$, while grouping the terms according to increasing powers of $N^{-\frac{1}{2}}$ gives another asymptotic expansion called the *Edgeworth series* of $\mathcal{P}$. Both expansions comprise a leading Gaussian term which embodies the *central limit theorem*:

$$\mathcal{P}(\boldsymbol{E}) \approx \frac{1}{\sqrt{\det(2\pi\mathbf{Q})}} \exp\bigl(-\tfrac{1}{2} \boldsymbol{E}^{\mathrm{T}} \mathbf{Q}^{-1} \boldsymbol{E}\bigr), \qquad \text{where} \qquad \boldsymbol{E} = \frac{\boldsymbol{X} - \langle \boldsymbol{X} \rangle}{\sqrt{N}}. \tag{8.30}$$

8.3.6 The saddlepoint approximation

A limitation of the Edgeworth series is that it gives an accurate estimate of $\mathcal{P}(\boldsymbol{X})$ only in the vicinity of $\boldsymbol{X} = \langle \boldsymbol{X} \rangle$, that is, for small values of $\boldsymbol{E}$. These convergence difficulties are easily understood: one is substituting a *local* approximation to $\log M$ (viz., a Taylor series expansion valid near $\boldsymbol{t} = \mathbf{0}$) into an integral, whereas integration is a *global* process which consults values of $\log M$ far from $\boldsymbol{t} = \mathbf{0}$. More specifically, under the simplifying assumption made earlier that we are looking at a centred distribution (that is, $\langle \boldsymbol{X} \rangle = \mathbf{0}$), the exact expression (8.23) of $\mathcal{P}(\boldsymbol{X})$ as the Fourier transform of $C^N(\boldsymbol{t})$ at $\boldsymbol{X}$ makes it intuitively clear that expanding $C(\boldsymbol{t})$ at $\boldsymbol{t} = \mathbf{0}$ (where it has its maximum modulus of one) will lead to a good approximation of the *integral* of $C^N(\boldsymbol{t})$, viz. $\mathcal{P}(\mathbf{0})$, or of its low-order frequency components, viz. $\mathcal{P}(\boldsymbol{X})$ for $\boldsymbol{X}$ near $\mathbf{0}$, but will fail to represent adequately the rapid oscillations of $C^N(\boldsymbol{t})$ which contribute to $\mathcal{P}(\boldsymbol{X})$ for $\boldsymbol{X}$ far from $\mathbf{0}$.

It is possible, however, to let the point $\boldsymbol{t}$ where $\log M$ is expanded as a Taylor series depend on the particular value $\boldsymbol{X}^*$ of $\boldsymbol{X}$ for which an accurate evaluation of $\mathcal{P}(\boldsymbol{X})$ is desired, by means of (8.24). By the previous reasoning, this will require identifying the point $\boldsymbol{t}$ where $C^N(\boldsymbol{t})$ oscillates as $\exp(+\mathrm{i}\boldsymbol{t} \cdot \boldsymbol{X}^*)$ and hence whose vicinity will contribute most to the integral representing $\mathcal{P}(\boldsymbol{X}^*)$. This is the essence of the *saddlepoint method* (Fowler 1936; Khinchin 1949; Daniels 1954; de Bruijn 1970; Bleistein and Handelsman 1986), which uses an analytical continuation of $M(\boldsymbol{t})$ from a function over $\mathbb{R}^n$ to a function over $\mathbb{C}^n$ (see, for example, Bricogne (1984, Section 5.1)). Putting, $\boldsymbol{t} = \boldsymbol{s} - \mathrm{i}\boldsymbol{\tau}$, the $\mathbb{C}^n$-version of Cauchy's theorem (Hörmander 1973) gives rise to the identity

$$\mathcal{P}(\boldsymbol{X}^*) = \frac{\exp(-\boldsymbol{\tau} \cdot \boldsymbol{X}^*)}{(2\pi)^n} \int_{\mathbb{R}^n} \exp\Big(N\big(\log M(\boldsymbol{\tau} + \mathrm{i}\boldsymbol{s}) - \mathrm{i}\boldsymbol{s} \cdot \boldsymbol{X}^*/N\big)\Big)\, \mathrm{d}^n\boldsymbol{s}, \tag{8.31}$$

for any $\boldsymbol{\tau} \in \mathbb{R}^n$. By a convexity argument involving the positive-definiteness of the covariance matrix $\mathbf{Q}$, there is a unique value of $\boldsymbol{\tau}$ such that

$$\nabla(\log M)\big|_{\boldsymbol{t}=\mathbf{0}-\mathrm{i}\boldsymbol{\tau}} = \frac{\boldsymbol{X}^*}{N}. \tag{8.32}$$

At the saddlepoint $\boldsymbol{t}^* = \mathbf{0} - \mathrm{i}\boldsymbol{\tau}$, the modulus of the integrand above is a maximum and its phase is stationary with respect to the integration variable $\boldsymbol{s}$: as N tends to infinity, all contributions to the integral cancel because of rapid oscillation, except those coming from the immediate vicinity of $\boldsymbol{t}^*$ where there is no oscillation. A Taylor expansion of $\log M^N$ to second order with respect to $\boldsymbol{s}$ at $\boldsymbol{t}^*$ then gives

$$\log M^N(\boldsymbol{\tau} + \mathrm{i}\boldsymbol{s}) \approx \log M^N(\boldsymbol{\tau}) + \mathrm{i}\boldsymbol{s} \cdot \boldsymbol{X}^* - \tfrac{1}{2}N\boldsymbol{s}^{\mathrm{T}}\mathbf{Q}\boldsymbol{s} \tag{8.33}$$

and hence

$$\mathcal{P}(\boldsymbol{X}^*) \approx \exp\bigl(\log M^N(\boldsymbol{\tau}) - \boldsymbol{\tau}\cdot\boldsymbol{X}^*\bigr) \times \frac{1}{(2\pi)^n}\int_{\mathbb{R}^n} \exp\bigl(-\tfrac{1}{2}\boldsymbol{s}^{\mathrm{T}}\mathbf{Q}'\boldsymbol{s}\bigr)\,\mathrm{d}^n\boldsymbol{s}.$$

The last integral is elementary and gives the 'saddlepoint approximation':

$$\mathcal{P}^{\mathrm{SP}}(\boldsymbol{X}^*) = \frac{\exp(\mathcal{S})}{\sqrt{\det(2\pi\mathbf{Q}')}}, \tag{8.34}$$

where

$$\mathcal{S} = \log M^N(\boldsymbol{\tau}) - \boldsymbol{\tau}\cdot\boldsymbol{X}^* \tag{8.35}$$

and where

$$\mathbf{Q}' = \nabla\nabla^{\mathrm{T}}(\log M^N) = N\mathbf{Q}. \tag{8.36}$$

This approximation scheme amounts to using the 'conjugate distribution' (Khinchin 1949)

$$P_{\boldsymbol{\tau}}(\boldsymbol{X}_j) = P(\boldsymbol{X}_j)\frac{\exp(\boldsymbol{\tau}\cdot\boldsymbol{X}_j)}{M(\boldsymbol{\tau})}, \tag{8.37}$$

instead of the original distribution $P(\boldsymbol{X}_j)$, which corresponds to $\boldsymbol{\tau} = \mathbf{0}$, for the common distribution of all N random vectors $\boldsymbol{X}_j$. The exponential modulation results from the analytic continuation of the characteristic (or moment-generating) function into $\mathbb{C}^n$ (Bricogne 1984, Section 5.2).

8.4 The statistical theory of phase determination

The methods of probability theory just surveyed were applied to various problems formally similar to the crystallographic phase problem (for example, the 'problem of the random walk' of Pearson (1905)) by Rayleigh (1880, 1905, 1918, 1919) and Kluyver (1905). They became the basis of the statistical theory of communication with the classic papers of Rice (1944, 1945).

The Gram–Charlier and Edgeworth series were introduced into crystallography by Bertaut (1955*a*,*b*,*c*, 1956*a*) and by Klug (1958) respectively, who showed them to constitute the mathematical basis of numerous formulae derived by Hauptman and Karle (1953). The saddlepoint approximation was introduced in this context by the present writer (Bricogne 1984).

8.4.1 Vectors of trigonometric structure factor expressions

Consider a crystal with given lattice and space-group symmetry G. Let H be a collection of reflections $\mathbf{h} \neq \mathbf{0}$ which are *symmetry-unique*, that is, no two of them are related by the action of G or by central inversion, and let D be an *asymmetric unit*, that is, a region of the unit cell no two points of which are equivalent under the action of G. The contributions to the vector $\boldsymbol{F}$ of structure factor values for reflections $\mathbf{h} \in H$, of a point atom of unit weight placed at position $\boldsymbol{x}$ in D may be arranged into a column vector $\boldsymbol{\xi}(\boldsymbol{x})$, whose total dimension over the reals depends on the number and the symmetry characteristics of the reflections in H (for full details, see, for example, Bricogne (1988*a*, Section 0.0 and Section 0.1)). Each coordinate $\xi_1(\boldsymbol{x}), \xi_2(\boldsymbol{x}), \ldots, \xi_n(\boldsymbol{x})$ of this vector is a function of $\boldsymbol{x}$ (a 'trigonometric structure factor expression') which is completely specified by knowledge of H and G.

8.4.2 Probability distributions of random atoms and moment-generating functions

Suppose now that position $\boldsymbol{x}$ in D is a random vector with probability density $m(\boldsymbol{x})$. Then $\boldsymbol{\xi}(\boldsymbol{x})$ is itself a random vector in $\mathbb{R}^n$, whose distribution $p(\boldsymbol{\xi})$ is the image of distribution $m(\boldsymbol{x})$ under the mapping $\boldsymbol{x} \rightarrow \boldsymbol{\xi}(\boldsymbol{x})$ just defined. The locus of $\boldsymbol{\xi}(\boldsymbol{x})$ in $\mathbb{R}^n$ is a compact algebraic manifold $\mathcal{L}$ (the multidimensional analogue of a Lissajous curve), so that p is a singular measure concentrated on that manifold, hence with compact support. The average with respect to p of any function Ω over $\mathbb{R}^n$ which is infinitely differentiable in a neighbourhood of $\mathcal{L}$ may be calculated as an average with respect to m over D by the 'induction formula'

$$\langle p, \Omega \rangle = \int_D m(\boldsymbol{x}) \Omega\big(\boldsymbol{\xi}(\boldsymbol{x})\big) \, \mathrm{d}^3\boldsymbol{x}. \tag{8.38}$$

In particular, one can calculate the moment-generating function M for distribution p as

$$M(\boldsymbol{t}) \equiv \big\langle p_\xi, \exp(\boldsymbol{t} \cdot \boldsymbol{\xi}) \big\rangle = \int_D m(\boldsymbol{x}) \exp\big(\boldsymbol{t} \cdot \boldsymbol{\xi}(\boldsymbol{x})\big) \, \mathrm{d}^3\boldsymbol{x} \tag{8.39}$$

If this expression is analytically tractable, we can calculate the moments μ of p, for which (8.38) gives the definition

$$\mu_{r_1, r_2, \ldots, r_n} \equiv \int_D m(\boldsymbol{x}) \xi_1^{r_1}(\boldsymbol{x}) \xi_2^{r_2}(\boldsymbol{x}) \cdots \xi_n^{r_n}(\boldsymbol{x}) \, \mathrm{d}^3\boldsymbol{x} \tag{8.40}$$

by differentiation of M according to (8.18). Alternatively, we can find the cumulants κ from $\log M$, as in (8.20). Conversely, the *structure factor algebra* for group G (Bertaut 1955*c*, 1956*b,c*, 1959*a,b*; Bertaut and Waser 1957) allows one to express products of ξ's in (8.40) as linear combinations of ξ's belonging to other reflections, and hence to calculate directly all moments and cumulants of the distribution $p(\boldsymbol{\xi})$ as linear combinations of real and imaginary parts of the Fourier coefficients of the prior distribution of atoms $m(\boldsymbol{x})$. The numerical values of these μ's and κ's can then be used to construct the first few terms of the Taylor series of M and $\log M$ respectively. This procedure plays a key role in the use of non-uniform probability distributions of atoms (see comment following (8.67)).

8.4.3 The joint probability distribution of structure factors

In the random atom model of an equal-atom structure, N atoms are placed randomly, independently of each other, in the asymmetric unit cell D of the crystal with probability density $m(\boldsymbol{x})$. For point atoms of unit weight (scattering factors other than unity can easily be introduced (Bricogne 1988*a*)), the vector $\boldsymbol{F}$ of structure factor values for reflections $\mathbf{h} \in H$ may be written

$$\boldsymbol{F} = \sum_{I=1}^{N} \boldsymbol{\xi}^{[I]}, \tag{8.41}$$

where the N versions $\boldsymbol{\xi}^{[I]}$ of the random vector $\boldsymbol{\xi}$ are independent and have the same distribution $p(\boldsymbol{\xi})$.

The joint probability distribution $\mathcal{P}(\boldsymbol{F})$ is then (Section 8.3.5)

$$\mathcal{P}(\boldsymbol{F}) = \frac{1}{(2\pi)^n} \int_{\mathbb{R}^n} \exp\bigl(N \log M(\mathrm{i}\boldsymbol{t}) - \mathrm{i}\boldsymbol{t} \cdot \boldsymbol{F}\bigr)\, \mathrm{d}^n\boldsymbol{t}, \tag{8.42}$$

and one of the approximation schemes described in Sections 8.3.5 and 8.3.6 must be used to evaluate it.

8.4.4 Overcoming the limitations of the Edgeworth series

The asymptotic expansions (8.29) of Gram–Charlier and Edgeworth have reasonable properties only if $F_{\mathbf{h}}$ lies in the vicinity of $\langle F_{\mathbf{h}} \rangle$ for all $\mathbf{h} \in H$ (see the discussion at the beginning of Section 8.3.6). Previous work on the j.p.d. of structure factors has used for $m(\boldsymbol{x})$ a uniform distribution, so that $\langle \boldsymbol{F} \rangle = \mathbf{0}$; as a result, the corresponding expansions are accurate only if all moduli $|F_{\mathbf{h}}|$ are small, in which case the j.p.d. contains little phase information. As was first pointed by Bricogne (1984), conventional direct methods lose sight of the fact that the Edgeworth series is a functional expression suitable only for local approximation of $\mathcal{P}(\boldsymbol{F})$ near $\boldsymbol{F} = \mathbf{0}$ and mistakenly assume it to be the global form for that function 'in the large'.

These two observations show that the traditional form of direct methods is a rather rudimentary implementation of the ideas of probability theory. Fortunately, better analytical devices can be found which circumvent these difficulties; the reader is referred to Bricogne (1984) for a full exposition, which will be freely summarized here. The main idea is that, if the locus $\mathcal{T}$ (a high-dimensional torus) defined by the large moduli is too extended in structure factor space for a single asymptotic expansion of $\mathcal{P}(\boldsymbol{F})$ to be accurate everywhere on it, then we should break $\mathcal{T}$ up into subregions, and construct a different local approximations to $\mathcal{P}(\boldsymbol{F})$ in each of them. Each of these subregions will consist of a 'patch' of $\mathcal{T}$, surrounding a point $\boldsymbol{F}^* \neq \mathbf{0}$ located on $\mathcal{T}$. Such a point $\boldsymbol{F}^*$ is obtained by assigning 'trial' phase values to the known moduli, but these trial values do not have to be viewed as 'serious' assumptions concerning the true values of the phases: rather, they should be thought of as pointing to a patch of $\mathcal{T}$ and to a specialized asymptotic expansion of $\mathcal{P}(\boldsymbol{F})$ designed to be the most accurate approximation possible to $\mathcal{P}(\boldsymbol{F})$ on that patch. With a sufficiently rich collection of such constructs, $\mathcal{P}(\boldsymbol{F})$ can be accurately calculated anywhere on $\mathcal{T}$.

We are thus led inevitably to the notion of *recentring*. Recentring the usual Gram–Charlier or Edgeworth asymptotic expansion for $\mathcal{P}(\boldsymbol{F})$ away from $\boldsymbol{F} = \mathbf{0}$, by making trial phase assignments which define a point $\boldsymbol{F}^*$ on $\mathcal{T}$, is equivalent to using a non-uniform probability distribution $q(\boldsymbol{x})$.

The latter constraint, however, leaves $q(\boldsymbol{x})$ highly indeterminate, but it turns out that there is a uniquely defined 'best' choice for it: it is that distribution which has *maximum entropy* under these constraints, entropy being a measure of closeness to uniformity. This result can be justified in two ways (Bricogne 1984, Sections 3 and 5).

1. The first way invokes Shannon's theory of information (Shannon and Weaver 1949), according to which any decrease in the entropy of the probability distribution of atoms $q(\boldsymbol{x})$ tends to diminish the size of the statistical ensemble of structures which can be generated using $q(\boldsymbol{x})$; then it invokes Jaynes' heuristic principle (Jaynes 1957) that one should rule out as few structures as possible on the basis of the limited information available in $\boldsymbol{F}^*$, to conclude that $q(\boldsymbol{x})$ should have maximum entropy (call it $q^{\mathrm{ME}}(\boldsymbol{x})$).
2. The second way relies on the saddlepoint approximation described in Section 8.3.6. The remarkable thing is that, without ever mentioning entropy nor invoking any semi-philosophical arguments, it yields an identical expression for the best choice of a non-uniform $q(\boldsymbol{x})$. The saddlepoint approximation to any conditional distribution conditioned on the knowledge of $\boldsymbol{F}^*$ is then the Gaussian approximation built from $q^{\mathrm{ME}}(\boldsymbol{x})$.

These two approaches, and the equivalence between them, will now be described.

8.5 The maximum entropy method

8.5.1 The context of information theory

The standard scheme of direct methods uses as its starting point a source of random atomic positions. The picture of this process may be discretized by partitioning the asymmetric unit of the crystal into B equal boxes labelled 1 to B. Then the symbols used by the source are the labels of these boxes, in the sense that drawing symbol i is equivalent to placing an atom in box i. Sequences of such randomly produced abstract symbols are called *messages*; in our case, a message of length N is an N-atom random structure.

Any such discrete stochastic process, considered as a discrete source of symbols, generates not only a set but an *ensemble* of messages, that is, in Wiener's terms 'a repertory of possible messages, and over this repertory a measure determining the probability of these messages' (Wiener 1949). In our case, this probability measure reflects the fact that different random structures—and hence different sets of structure factors—will occur with different frequencies (or 'statistical weights') in the ensemble.

8.5.2 Definition and meaning of entropy

An important numerical quantity associated with a discrete source is its *entropy per symbol* H, which gives a measure of the 'amount of uncertainty' involved in the choice of a symbol. Suppose that successive symbols are independent and that symbol i has probability q_i. Then the general requirements that H should be a continuous function of the q_i, should increase with increasing uncertainty, and should be additive for independent sources of uncertainty, suffice to define H uniquely as

$$H(q_1, \ldots, q_n) = -k \sum_{i=1}^{n} q_i \log q_i, \tag{8.43}$$

where k is an arbitrary positive constant (Shannon and Weaver 1949, Appendix 2) whose value depends on the unit of entropy chosen.

Two important theorems (Shannon and Weaver 1949, Appendix 3) provide a more intuitive grasp of the meaning and importance of entropy:

1. H is approximately the logarithm of the reciprocal probability of a typical long sequence, divided by the number of symbols in the sequence;

2. H gives the rate of growth, with increasing sequence length, of the logarithm of the number of reasonably probable sequences, regardless of the precise meaning given to the criterion of being 'reasonably probable'.

The entropy H of a source is thus a direct measure of the strength of the restrictions placed on the permissible messages, greater restrictions leading to lower entropy. In our case, its maximum value $H_{\max} = \log B$ is reached when all the symbols are equally probable, that is, for a uniform prior distribution of the atoms. When the prior distribution is not uniform, the usage of the different symbols is biased away from this maximum freedom, and the entropy of the source is lower; by Shannon's second theorem, the number of 'reasonably probable' messages of a given length is decreased accordingly.

8.5.3 The maximum entropy criterion

We are now in a position to address the problem posed in Section 8.4.4, namely the choice of a non-uniform probability distribution of atoms $q(\boldsymbol{x})$ from the assumed knowledge of a limited set of its Fourier coefficients.

By the two theorems of Shannon quoted above, the entropy H of the source of random atomic positions defined by $q(\boldsymbol{x})$ affords a quantitative measure of the extent to which the range of random structures which can be generated with a reasonable probability has been narrowed down. Any reduction of the entropy of the source beyond that strictly necessary to accommodate the assumed knowledge will be reflected by a decrease in the number of reasonably probable random structures, and hence by an unjustified restriction to a subset of all possible structures consistent with the given data; this is equivalent to imposing extra constraints, not warranted by the data.

If the assumed knowledge of some structure factors $\boldsymbol{F}^*$ is to be reflected in a non-uniform probability distribution of atoms, the previous reasoning leads uniquely to choosing for $q(\boldsymbol{x})$ the maximum entropy distribution $q^{\mathrm{ME}}(\boldsymbol{x})$ having the corresponding values $\boldsymbol{U}^*$ of the unitary structure factors for its Fourier coefficients, since the latter defines the source which reproduces the assumed data with minimum bias. This is a particular instance of the maximum entropy principle of Jaynes (1957, 1968, 1983) which may be loosely phrased as follows:

> 'The probability assignment shall be the one with maximum entropy consistent with the available knowledge, so as to remain maximally non-committal with regard to missing information.'

The quantity which measures most directly the strength of the restrictions introduced by the non-uniformity of $q(\boldsymbol{x})$ is not the source entropy $H(q)$ itself, but rather the difference $H(q) - H_{\max}$ since the proportion of N-atom random structures which remain 'reasonably probable' in the ensemble of the corresponding source is $\exp\left(N\left[H(q)-H_{\max}\right]\right)$. This difference may be written (using continuous rather than discrete distributions) as

$$H(q) - H_{\max} = -\int_D q(\boldsymbol{x}) \log\left(\frac{q(\boldsymbol{x})}{m(\boldsymbol{x})}\right) \mathrm{d}^3\boldsymbol{x}, \tag{8.44}$$

where $m(\boldsymbol{x}) = \mathrm{vol}(D)^{-1}$ is the uniform distribution which is such that $H(m) = H_{\max} = \log\left(\mathrm{vol}(D)\right)$.

The final form of the maximum entropy criterion is thus that $q(\boldsymbol{x})$ should be chosen so as to maximize, under the constraints expressing the knowledge of some of its Fourier coefficients, its entropy

$$\mathcal{S}_m(q) = -\int_D q(\boldsymbol{x}) \log\left(\frac{q(\boldsymbol{x})}{m(\boldsymbol{x})}\right) \mathrm{d}^3\boldsymbol{x} \tag{8.45}$$

relative to the 'prior prejudice' $m(\boldsymbol{x})$ which maximizes H in the absence of such knowledge.

8.5.4 The maximum entropy formalism

Jaynes (1957) solved the problem of explicitly determining such maximum entropy distributions in the case of general linear constraints, using an analytical apparatus first exploited by Gibbs in statistical mechanics.

The maximum entropy distribution $q^{\mathrm{ME}}(\boldsymbol{x})$, under the prior prejudice $m(\boldsymbol{x})$, satisfying the linear constraint equations

$$\mathcal{C}_j(q) \equiv \int_D q(\boldsymbol{x}) C_j(\boldsymbol{x}) \, \mathrm{d}^3\boldsymbol{x} = c_j \qquad \text{with} \qquad j = 1, 2, \ldots, M, \tag{8.46}$$

where the $\mathcal{C}_j(q)$ are linear *constraint functionals* defined by given *constraint functions* $C_j(\boldsymbol{x})$, and the c_j are given *constraint values*, is obtained by maximizing with respect to q the relative entropy defined by (8.45). An extra constraint is the normalization condition

$$\mathcal{C}_0(q) \equiv \int_D q(\boldsymbol{x}) \, \mathrm{d}^3\boldsymbol{x} = 1, \tag{8.47}$$

to which it is convenient to give the label $j = 0$, so that it can be handled together with the others by putting $C_0(\boldsymbol{x}) = 1$ and $c_0 = 1$.

By a standard variational argument, in constant use in statistical mechanics, this constrained maximization is equivalent to the unconstrained maximization of the functional

$$\mathcal{S}_m(q) + \sum_{j=0}^{M} \lambda_j \mathcal{C}_j(q), \tag{8.48}$$

where the λ_j are Lagrange multipliers whose values may be determined from the constraints. This new variational problem is readily solved: if $q(\boldsymbol{x})$ is varied to $q(\boldsymbol{x}) + \delta q(\boldsymbol{x})$, the resulting variations in the functionals $\mathcal{S}_m$ and $\mathcal{C}_j$ will be

$$\begin{aligned} \delta\mathcal{S}_m &= \int_D \left(-1 - \log\frac{q(\boldsymbol{x})}{m(\boldsymbol{x})}\right)\delta q(\boldsymbol{x})\,\mathrm{d}^3\boldsymbol{x} \qquad \text{and} \\ \delta\mathcal{C}_j &= \int_D C_j(\boldsymbol{x})\delta q(\boldsymbol{x})\,\mathrm{d}^3\boldsymbol{x}, \end{aligned} \tag{8.49}$$

respectively. If the variation of the functional (8.48) is to vanish for arbitrary variations $\delta q(\boldsymbol{x})$, the integrand in the expression for that variation from (8.49) must vanish identically. Therefore the maximum entropy probability density distribution $q^{\mathrm{ME}}(\boldsymbol{x})$ satisfies the relation

$$-1 - \log\left(\frac{q(\boldsymbol{x})}{m(\boldsymbol{x})}\right) + \sum_{j=0}^{M} \lambda_j C_j(\boldsymbol{x}) = 0 \tag{8.50}$$

and hence

$$q^{\mathrm{ME}}(\boldsymbol{x}) = m(\boldsymbol{x}) \exp(\lambda_0 - 1) \exp\left(\sum_{j=1}^{M} \lambda_j C_j(\boldsymbol{x})\right). \tag{8.51}$$

It is convenient now to separate the multiplier λ_0 associated with the normalization constraint by putting

$$\lambda_0 - 1 = -\log Z, \tag{8.52}$$

where Z is a function of the other multipliers $\lambda_1, \ldots, \lambda_M$. The final expression for $q^{\mathrm{ME}}(\boldsymbol{x})$ is thus

$$q^{\mathrm{ME}}(\boldsymbol{x}) = \frac{m(\boldsymbol{x})}{Z(\lambda_1, \ldots, \lambda_M)} \exp\left(\sum_{j=1}^{M} \lambda_j C_j(\boldsymbol{x})\right). \tag{8.53}$$

The values of Z and of $\lambda_1, \ldots, \lambda_M$ may now be determined from the initial constraint equations. The normalization condition demands that

$$Z(\lambda_1, \ldots, \lambda_M) = \int_D m(\boldsymbol{x}) \exp\left(\sum_{j=1}^{M} \lambda_j C_j(\boldsymbol{x})\right) \mathrm{d}^3\boldsymbol{x}. \tag{8.54}$$

The generic constraint equations (8.46) determine $\lambda_1, \ldots, \lambda_M$ by the conditions that

$$\int_D \frac{m(\boldsymbol{x})}{Z} \exp\left(\sum_{k=1}^{M} \lambda_k C_k(\boldsymbol{x})\right) C_j(\boldsymbol{x})\,\mathrm{d}^3\boldsymbol{x} = c_j, \tag{8.55}$$

for $j = 1, 2, \ldots, M$. But, by Leibniz's rule of differentiation under the integral sign, these equations may be written in the compact form

$$\frac{\partial(\log Z)}{\partial\lambda_j} = c_j \qquad \text{with} \qquad j = 1, 2, \ldots, M. \tag{8.56}$$

Equations (8.53), (8.54) and (8.56) constitute the *maximum entropy equations.*

The maximum value attained by the entropy is readily found:

$$\begin{aligned}
\mathcal{S}_m(q^{\mathrm{ME}}) &= -\int_D q^{\mathrm{ME}}(\boldsymbol{x}) \log\left(\frac{q^{\mathrm{ME}}(\boldsymbol{x})}{m(\boldsymbol{x})}\right) \mathrm{d}^3\boldsymbol{x} \\
&= -\int_D q^{\mathrm{ME}}(\boldsymbol{x}) \left(-\log Z + \sum_{j=1}^{M} \lambda_j C_j(\boldsymbol{x})\right) \mathrm{d}^3\boldsymbol{x},
\end{aligned}$$

which is, using the constraint equations,

$$\mathcal{S}_m(q^{\mathrm{ME}}) = \log Z - \sum_{j=1}^{M} \lambda_j c_j. \tag{8.57}$$

The latter expression may be rewritten, by means of equations (8.56), as

$$\mathcal{S}_m(q^{\mathrm{ME}}) = \log Z - \sum_{j=1}^{M} \lambda_j \frac{\partial(\log Z)}{\partial\lambda_j}, \tag{8.58}$$

which shows that, in their dependence on the λ's, the entropy and $\log Z$ are related by Legendre duality.

Jaynes' maximum entropy theory also gives an estimate for $\mathcal{P}(\boldsymbol{F}^*)$ as a consequence of Shannon's second theorem (Section 8.5.2)

$$\mathcal{P}^{\mathrm{ME}}(\boldsymbol{F}^*) = \exp(\mathcal{S}), \tag{8.59}$$

where

$$\mathcal{S} = \log Z^N - \boldsymbol{\lambda} \cdot \boldsymbol{F}^* = N\mathcal{S}_m(q^{\mathrm{ME}}) \tag{8.60}$$

is the total entropy.

8.5.5 The crystallographic maximum entropy formalism

It is possible to solve the maximum entropy equations explicitly for the crystallographic case which has motivated this study, that is, for the purpose of constructing $q^{\mathrm{ME}}(\boldsymbol{x})$ from the knowledge of a set of trial structure-factor values. These derivations are given by Bricogne (1984, Sections 3.4 and 3.5). Extensive relations with the algebraic formalism of traditional direct methods are exhibited in Section 4, and connections with the theory of determinantal inequalities and with the maximum determinant rule of Tsoucaris (1970) are studied in Section 6 of the same paper, which the reader interested in these topics is invited to consult.

8.6 The saddlepoint method

The saddlepoint method (Section 8.3.6) constitutes an alternative approach to the problem of evaluating the joint probability $\mathcal{P}(\boldsymbol{F}^*)$ of structure factors when some of the moduli in $\boldsymbol{F}^*$ are large. As shown previously this approximation amounts to using the 'conjugate distribution'

$$p_{\boldsymbol{\tau}}(\boldsymbol{\xi}) = p(\boldsymbol{\xi}) \frac{\exp(\boldsymbol{\tau} \cdot \boldsymbol{\xi})}{M(\boldsymbol{\tau})}, \tag{8.61}$$

instead of the original distribution $p(\boldsymbol{\xi}) = p_{\mathbf{0}}(\boldsymbol{\xi})$ for the distribution of random vector $\boldsymbol{\xi}$. This conjugate distribution $p_{\boldsymbol{\tau}}$ is induced from the modified probability distribution of atoms

$$q_{\boldsymbol{\tau}}(\boldsymbol{x}) = m(\boldsymbol{x}) \frac{\exp\big(\boldsymbol{\tau} \cdot \boldsymbol{\xi}(\boldsymbol{x})\big)}{M(\boldsymbol{\tau})}, \tag{8.62}$$

where, by the induction formula (8.38), $M(\boldsymbol{\tau})$ may be written as

$$M(\boldsymbol{\tau}) = \int_D m(\boldsymbol{x}) \exp\big(\boldsymbol{\tau} \cdot \boldsymbol{\xi}(\boldsymbol{x})\big)\, \mathrm{d}^3\boldsymbol{x} \tag{8.63}$$

and where $\boldsymbol{\tau}$ is the unique solution of the saddlepoint equation

$$\nabla_{\boldsymbol{\tau}}(\log M^N) = \boldsymbol{F}^*. \tag{8.64}$$

The desired approximation is then

$$\mathcal{P}^{\mathrm{SP}}(\boldsymbol{F}^*) = \frac{\exp(\mathcal{S})}{\sqrt{\det(2\pi\mathbf{Q}')}}, \tag{8.65}$$

where

$$\mathcal{S} = \log M^N(\boldsymbol{\tau}) - \boldsymbol{\tau} \cdot \boldsymbol{F}^* \tag{8.66}$$

and where

$$\mathbf{Q}' = \nabla\nabla^{\mathrm{T}}(\log M^N) = N\mathbf{Q}. \tag{8.67}$$

Note that the elements of the Hessian matrix $\mathbf{Q} = \nabla\nabla^{\mathrm{T}}(\log M)$, which may be used to solve equation (8.64) for $\boldsymbol{\tau}$ by Newton's method, are just the trigonometric second-order cumulants of distribution p, and hence can be calculated *via* structure factor algebra from the Fourier coefficients of $q_{\boldsymbol{\tau}}(\boldsymbol{x})$. All the quantities involved in the expression for $\mathcal{P}^{\mathrm{SP}}(\boldsymbol{F}^*)$ are therefore effectively computable from the initial data, $m(\boldsymbol{x})$ and $\boldsymbol{F}^*$.

8.6.1 Relation between the maximum entropy method and the saddlepoint method

Comparison of the three equations (8.62), (8.63) and (8.64) with (8.53), (8.54) and (8.56) shows complete identity, up to an obvious change in notation:

- the Z-function in (8.54) is the M-function in (8.63);
- the Lagrange multipliers in (8.56) are the coordinates of the saddlepoint in (8.64).

There is thus complete equivalence between the maximum entropy approach to the phase problem and the classical probabilistic approach by the method of joint distributions, provided the latter is enhanced by the adoption of the saddlepoint approximation.

This rather striking result is not without precedent. The analytic proof of the maximum entropy principle given here is an exact parallel of the Darwin–Fowler formulation of statistical mechanics (Fowler 1936), which also bypasses the established route—explicit combinatorial enumerations followed by an appeal to Stirling's formula—by means of a direct saddlepoint approximation to the thermodynamic partition function. Both results may ultimately be rationalized by recalling that Stirling's formula itself can be established by a saddlepoint approximation to Euler's Γ-function (de Bruijn 1970).

The saddlepoint method is actually slightly superior to the maximum entropy method. Comparing (8.59) and (8.60) on the one hand with (8.65) and (8.66) on the other hand, we see that $\mathcal{P}^{\mathrm{ME}}$ is almost identical to $\mathcal{P}^{\mathrm{SP}}$, but lacks the denominator. The latter, which is the normalization factor of a multivariate Gaussian with covariance matrix $\mathbf{Q}'$ as given in (8.67), may easily be seen to arise from the extra logarithmic term in Stirling's formula:

$$\log(q!) \approx q \log q - q + \tfrac{1}{2}\log(2\pi q) \tag{8.68}$$

(see, for instance, Lebedev 1972) beyond the first two terms which serve to define entropy, since by Szegö's theorem (Szegö 1920)

$$\frac{1}{n}\log\det(2\pi\mathbf{Q}) \approx \int_D \log\left(2\pi q^{\mathrm{ME}}(\boldsymbol{x})\right)\,\mathrm{d}^3\boldsymbol{x}. \tag{8.69}$$

The relative effect of this extra normalization factor depends on the ratio

$$\frac{n}{N} = \frac{\text{dimension of } \boldsymbol{F} \text{ over } R}{\text{number of atoms}}$$

and hence becomes predominant at high resolution.

This superiority of the saddlepoint approximation (8.65) over the standard maximum entropy result (8.59) was pointed out by Bricogne (1988*a*). Later, Gull (1989) arrived at a similar correction for the maximum entropy result, involving the logarithm of a determinant, by a different procedure. The need to append yet further arguments to those already invoked to justify the maximum entropy principle makes the directness and simplicity of the saddlepoint method stand out even more strikingly.

8.7 Summary

The power of conventional probabilistic direct methods of phase determination was limited by two factors:

1. the inability of the random atom model to represent the subtle correlations between atomic positions implied by the laws of chemistry;
2. any extraneous inadequacy in implementing the random atom model itself.

This study, first presented in Bricogne (1984), has been directed at the second problem, and has shown that severe degradation of the potential of the random atom model had resulted from the time-honoured recourse to the Edgeworth series for approximating joint probability distributions of structure factors.

Two remedies, at first sight very different from each other, have been examined: the maximum entropy method, and the saddlepoint approximation. They have been shown to be essentially equivalent. The reader may share the writer's relief at having replaced the invocation of an all-embracing epistemological argument (however sound and however profound) by the certainty of a purely analytical derivation, especially as the latter gave a better result until the former was revised.

The above relation between entropy maximization and the saddlepoint approximation is the basis of a Bayesian statistical approach to the phase problem (Bricogne 1988*a*) where the assumptions under which joint distributions of structure factors are sought incorporate many new ingredients (such as molecular boundaries, isomorphous substitutions, known fragments, non-crystallographic symmetries, multiple crystal forms) besides trial phase choices for basis reflections. The maximum entropy criterion intervenes in the construction of non-uniform probability distributions of random atomic positions under these assumptions, and these distributions continue to play the role of useful computational intermediates in obtaining the approximate joint probability $\mathcal{P}^{\mathrm{SP}}(\boldsymbol{F}^*)$ and the associated conditional distributions and likelihood functions.

The tree-directed multisolution approach alluded to in Section 8.2 has been implemented (Bricogne and Gilmore 1990) and has successfully solved small molecule structures (Gilmore *et al.* 1990). Recently, the expression of the likelihood criterion has been extended to deal with the reflection overlaps present in powder and Laue diffraction data (Bricogne 1991), and the same strategy applied with this new likelihood has successfully solved two moderately large inorganic crystal structures from powder data alone (Gilmore *et al.* 1991). Applications to macromolecular structures are under way. Woolfson's appraisal of the contribution of the maximum entropy approach to crystallography, quoted in Section 8.2.1, may thus soon require substantial revision.

References

Bertaut, E. F. (1955*a*). La méthode statistique en cristallographie. I. *Acta Crystallographica*, **8**, 537–43.

Bertaut, E. F. (1955*b*). La méthode statistique en cristallographie. II. Quelques applications. *Acta Crystallographica*, **8**, 544–8.

Bertaut, E. F. (1955*c*). Fonctions de répartition: application à l'approche directe des structures. *Acta Crystallographica*, **8**, 823–32.

Bertaut, E. F. (1956*a*). Les groupes de translation non primitifs et la méthode statistique. *Acta Crystallographica*, **9**, 322.

Bertaut, E. F. (1956*b*). Tables de linéarisation des produits et puissances des facteurs de structure. *Acta Crystallographica*, **9**, 322–3.

Bertaut, E. F. (1956*c*). Algèbre des facteurs de structure. *Acta Crystallographica*, **9**, 769–70.

Bertaut, E. F. (1959*a*). IV. Algèbre des facteurs de structure. *Acta Crystallographica*, **12**, 541–9.

Bertaut, E. F. (1959*b*). V. Algèbre des facteurs de structure. *Acta Crystallographica*, **12**, 570–4.

Bertaut, E. F. and Waser, J. (1957). Structure factor algebra. II. *Acta Crystallographica*, **10**, 606–7.

Bhattacharya, R. N. and Ranga Rao, R. (1976). *Normal approximation and asymptotic expansions.* Wiley, New York.

Bleistein, N. and Handelsman, R. A. (1986). *Asymptotic expansions of integrals.* Dover, New York.

Bricogne, G. (1984). Maximum entropy and the foundations of direct methods. *Acta Crystallographica*, **A40**, 410–45.

Bricogne, G. (1988*a*). A Bayesian statistical theory of the phase problem. I. A multichannel maximum-entropy formalism for constructing generalized joint probability distributions of structure factors. *Acta Crystallographica*, **A44**, 517–45.

Bricogne, G. (1988*b*). Maximum entropy methods in the X-ray phase problem. In *Crystallographic Computing 4* (ed. N. W. Isaacs and M. R. Taylor), pp. 60–79. Oxford University Press, New York.

Bricogne, G. (1991). A multisolution method of phase determination by combined maximization of entropy and likelihood. III. Extension to powder diffraction data. Submitted to *Acta Crystallographica.*

Bricogne, G. and Gilmore, C. J. (1990). A multisolution method of phase determination by combined maximization of entropy and likelihood. I. Theory, algorithms and strategy. *Acta Crystallographica*, **A46**, 284–97.

Bruijn, N. G. de (1970). *Asymptotic methods in analysis* (3rd edn). North-Holland, Amsterdam.

Cramér, H. (1946). *Mathematical methods of statistics.* Princeton University Press.

Daniels, H. E. (1954). Saddlepoint approximations in statistics. *Annals of Mathematical Statistics*, **25**, 631–50.

Fowler, R. H. (1936). *Statistical mechanics. The theory of the properties of matter in equilibrium* (2nd edn). Cambridge University Press.

Germain, G. and Woolfson, M. M. (1968). On the application of phase relationships to complex structures. *Acta Crystallographica*, **B24**, 91–6.

Giacovazzo, C. (1980). *Direct methods in crystallography.* Academic Press, London.

Gilmore, C. J., Bricogne, G., and Bannister, C. (1990). A multisolution method of phase determination by combined maximization of entropy and likelihood. II. Application to small molecules. *Acta Crystallographica*, **A46**, 297–308.

Gilmore, C. J., Henderson, K., and Bricogne, G. (1991). A multisolution method of phase determination by combined maximization of entropy and likelihood. IV. The *ab initio* solution of crystal structures from their X-ray powder diffraction data. Submitted to *Acta Crystallographica.*

Gull, S. F. (1989). Developments in maximum entropy data analysis. In *Maximum entropy and Bayesian methods, Cambridge, England, 1988* (ed. J. Skilling), pp. 53–71. Kluwer, Dordrecht.

Harker, D. and Kasper, J. S. (1948). Phases of Fourier coefficients directly from crystal diffraction data. *Acta Crystallographica*, **1**, 70–5.

Hauptman, H. (1972). *Crystal structure determination: the role of the cosine seminvariants.* Plenum Press, New York.

Hauptman, H. (1975). The joint probability distribution of seven structure factors. *Acta Crystallographica*, **A31**, 671–9.

Hauptman, H. (1980). Probabilistic theory of the structure seminvariants. In *Theory and practice of direct methods in crystallography* (ed. M. C. F. Ladd and R. A. Palmer), pp. 151–97. Plenum, New York.

Hauptman, H. and Karle, J. (1953). *The solution of the phase problem: I. The centrosymmetric crystal.* ACA Monograph No. 3. Polycrystal Book Service, Pittsburgh.

Hörmander, L. (1973). *An introduction to complex analysis in several variables* (2nd edn). North-Holland, Amsterdam.

Jaynes, E. T. (1957). Information theory and statistical mechanics. *Physical Review*, **106**, 620–30.

Jaynes, E. T. (1968). Prior probabilities. *IEEE Transactions on Systems Science and Cybernetics*, **SSC-4**, 227–41.

Jaynes, E. T. (1983). *Papers on probability, statistics and statistical physics*, Synthese Library, Vol. 158 (ed. R. D. Rosenkrantz). Reidel, Dordrecht.

Karle, J. and Hauptman, H. (1950). The phases and magnitudes of the structure factors. *Acta Crystallographica*, **3**, 181–7.

Karle, J. and Karle, I. L. (1966). The symbolic addition procedure for phase determination for centrosymmetric and noncentrosymmetric crystals. *Acta Crystallographica*, **21**, 849–59.

Khinchin, A. I. (1949). *Mathematical foundations of statistical mechanics*. Dover, New York.

Klug, A. (1958). Joint probability distributions of structure factors and the phase problem. *Acta Crystallographica*, **11**, 515–43.

Kluyver, J. C. (1905). A local probability problem. *Koninklijke Akademie van Wetenschappen te Amsterdam: Proceedings of the Section of Sciences*, **8**, 341–50.

Lebedev, N. N. (1972). *Special functions and their applications.* Dover, New York.

Lindley, D. V. (1965). *Introduction to probability and statistics from a Bayesian viewpoint. Part 1: Probability* and *Part 2: Inference.* Cambridge University Press.

Neyman, J. and Pearson, E. S. (1933). On the problem of the most efficient tests of statistical hypotheses. *Philosophical Transactions of the Royal Society of London*, **231A**, 289–337.

Nilsson, N. J. (1971). *Problem-solving methods in artificial intelligence.* McGraw-Hill, New York.

Paley, R. E. A. C. and Wiener, N. (1934). *Fourier transforms in the complex domain.* American Mathematical Society, Providence, Rhode Island.

Pearson, K. (1905). The problem of the random walk. *Nature*, **72**, 294 and 342.

Rayleigh, Lord (1880). On the resultant of a large number of vibrations of the same pitch and of arbitrary phase. *London, Edinburgh, and Dublin Philosophical Magazine and Journal of Science*, Fifth Series, **10**, 73–8.

Rayleigh, Lord (1905). The problem of the random walk. *Nature*, **72**, 318.

Rayleigh, Lord (1918). On the light emitted from a random distribution of luminous sources. *London, Edinburgh, and Dublin Philosophical Magazine and Journal of Science*, Sixth Series, **36**, 429–49.

Rayleigh, Lord (1919). On the problem of random vibrations, and of random flights in one, two or three dimensions. *London, Edinburgh, and Dublin Philosophical Magazine and Journal of Science*, Sixth Series, **37**, 321–47.

Rice, S. O. (1944, 1945). Mathematical analysis of random noise, I–IV. *Bell System Technical Journal*, **23**, 283–332 and **24**, 46–156. Reprinted in Wax, N. (ed.) (1954). *Selected papers on noise and stochastic processes*, pp. 133–294. Dover, New York.

Sayre, D. (1952). The squaring method: a new method for phase determination. *Acta Crystallographica*, **5**, 60–5.

Shannon, C. E. and Weaver, W. (1949). *The mathematical theory of communication.* University of Illinois Press, Urbana.

Szegö, G. (1920). Beiträge zur Theorie der Toeplitzschen Formen (Erste Mitteilung). *Mathematische Zeitschrift*, **6**, 167–202.

Thorpe, E. O. (1966). *Beat the dealer.* Vintage Books, New York.

Tsoucaris, G. (1970). A new method of phase determination. The 'maximum determinant rule'. *Acta Crystallographica*, **A26**, 492–9.

Wiener, N. (1949). *Extrapolation, interpolation and smoothing of stationary time series.* M.I.T. Press, Cambridge, Massachusetts.

Wilson, A. J. C. (1949). The probability distribution of X-ray intensities. *Acta Crystallographica*, **2**, 318–21.

Woolfson, M. M. (1987). Direct methods—from birth to maturity. *Acta Crystallographica*, **A43**, 593–612.

Index